Chemical Process Equipment: Selection and Design

Chemical Process Equipment: Selection and Design

Editor

Ashish Sharma

Chemical Process Equipment: Selection and Design

Edited by **Ashish Sharma**

Printed in 2017

ISBN: 978-1-68117-342-9

Library of Congress Control Number: 2015939254

© 2016 by
SCITUS Academics LLC,
616, Corporate Way, Suite 2, 4766,
Valley Cottage, NY 10989

www.scitusacademics.com

This book contains information obtained from highly regarded resources. Copyright for individual articles remains with the authors as indicated. All chapters are distributed under the terms of the Creative Commons Attribution License, which permits unrestricted use, distribution, and reproduction in any medium, provided the original author and source are credited.

Notice

Reasonable efforts have been made to publish reliable data and views articulated in the chapters are those of the individual contributors, and not necessarily those of the editors or publishers. Editors or publishers are not responsible for the accuracy of the information in the published chapters or consequences of their use. The publisher believes no responsibility for any damage or grievance to the persons or property arising out of the use of any materials, instructions, methods or thoughts in the book. The editors and the publisher have attempted to trace the copyright holders of all material reproduced in this publication and apologize to copyright holders if permission has not been obtained. If any copyright holder has not been acknowledged, please write to us so we may rectify.

Contents

Preface vii

Chapter 1   The Influence of Ceric Oxide on Phase Composition and Activity of Iron Oxide Catalysts 1

Aleksandr A. Lamberov, Ekaterina V. Dementyeva, Dmitriy I. Vavilov, Olga V. Kuzmina, Rinat R. Gilmullin, and Ekaterina A. Pavlova

Chapter 2   Kinetic Aspects of Gold and Silver Recovery in Cementation with Zinc Power and Electrocoagulation Iron Process 15

Gabriela V. Figueroa Martinez, José R. Parga Torres, Jesús L. Valenzuela García, Guillermo C. Tiburcio Munive, and Gregorio González Zamarripa

Chapter 3   Gasification Coupled Chemical Looping Combustion of Coal: A Thermodynamic Process Design Study 35

Sonali A. Borkhade, Preksha A. Shriwas, and Ganesh R. Kale

Chapter 4   Analysis of Changes in the Properties of Selected Chemical Compounds and Motor Fuels Taking Place during Oxidation Processes 65

Marta Skolniak, Paweł Bukrejewski, and Jarosław Frydrych

Chapter 5   Optimal Solutions of Multiproduct Batch Chemical Process Using Multiobjective Genetic Algorithm with Expert Decision System 109

Diab Mokeddem and Abdelhafid Khellaf

Chapter 6   Bio and Chemical Sensors Based on Surface Plasmon Resonance in a Plastic Optical Fiber 133

Nunzio Cennamo and Luigi Zeni

Chapter 7   A Methodology for Simultaneous Process and Product Design in the Formulated Consumer Products Industry: The Case Study of the Detergent Business 163

Mariano Martín and Alberto Martínez

Chapter 8	Advances in Pressure Swing Adsorption for Gas Separation..........205
	Carlos A. Grande

Citations..241
Index..245

Preface

Chemical Process Equipment is a results-oriented reference for engineers who specify, design, maintain or run chemical and process plants. This book delivers information on the selection, sizing and operation of process equipment in a format that enables quick and accurate decision making on standard process and equipment choices, saving time, improving productivity, and building understanding. Coverage emphasizes common real-world equipment design rather than experimental or esoteric and focuses on maximizing performance. Chemical process equipment is of two kinds: custom designed and built, or proprietary "off the shelf." For example, the sizes and performance of custom equipment such as distillation towers, drums, and heat exchangers are derived by the process engineer on the basis of established principles and data, although some mechanical details remain in accordance with safe practice codes and individual fabrication practices. The process design of proprietary equipment, as considered in this book, establishes its required performance and is a process of selection from the manufacturers' offerings, often with their recommendations or on the basis of individual experience.

Editor

The Influence of Ceric Oxide on Phase Composition and Activity of Iron Oxide Catalysts

Aleksandr A. Lamberov[1], Ekaterina V. Dementyeva[1], Dmitriy I. Vavilov[1], Olga V. Kuzmina[1], Rinat R. Gilmullin[2], and Ekaterina A. Pavlova[2]

[1]Department of Chemistry, Kazan Federal University, Kazan, Russia
[2]Scientifically-Technological Center, OJSC "Nizhnekamskneftekhim", Nizhnekamsk, Russia

ABSTRACT

The process of dehydrogenation of methyl butenes to isoprene is conducted in the presence of iron oxide catalysts whose composition may include oxides of alkaline metals, alkaline earth metals, and transition metals. Catalysts of latest generation can also contain oxides of rare earth elements, particularly cerium oxide. However there is no any common opinion concerning its effect on catalytic properties of iron oxide catalysts. It is well known that ceric oxide has a positive effect on the quantity and stability of active centers and can play a critical role in a redox cycle of the dehydrogenation process. By means of differential thermal analysis, dispersion analysis and X-ray phase analysis, it was found in present study that introducing of ceric oxide promotes the decrease in hematite crystallite sizes. At the same time, it prevents potassium polyferrites formation, with the equilibrium of topochemical reaction between ferric oxide and po tassium carbonates moving predominantly to the formation of intermediate products- monoferrite systems, having greater catalytic activity. The increase in potassium monoferrite content results in dispersion of particles in the Fe_2O_3- K_2CO_3-**CeO_2** system that is accompanied by modification of texture characteristics. For this catalyst composition, the optimum concentration of ceric oxide (8.7 wt %), leading to the formation of a certain ratio of monoand polyferrite phases, was found. If more than 8.7 wt% of **CeO_2** is introduced, the modification of texture characteristics of catalyst samples takes place that negatively affects their selectivity.

INTRODUCTION

As known from [1-3], the process of olefin hydrocarbons dehydrogenation is performed in presence of iron oxide catalysts where potassium ferrites are the active components that are formed in the process of solid phase interaction between ferric oxide and potassium compounds. Cerium compound which content varies in the range of 2 to 30 wt% [4, 5] is used as a promoter affecting the catalyst activity [6]. Ceric oxide is considered [6] to increase the number of active centers and to influence on their nature. Polyvalent cation (Ce^{4+}) of this metal has sufficient ionic strength intensifying the Fe-O link polarization that leads to increase in the basicity of catalyst surface where dehydrogenation reaction occurs [7]. In accordance with literature [7], cerium presence in catalyst promotes more intensive electron exchange between Fe^{2+} and Fe^{3+} ions which present in the polyferrite structure during dehydrogenation, thus preventing their complete recovery to Fe_3O_4. There is a few data in literature about ceric oxide effect on phase composition formation of iron oxide catalysts [6, 8, and 9].

The objective of this study is to define ceric oxide influence on phase composition, iron oxide catalysts activity and to find out the mechanism of its activity.

EXPERIMENTAL

Model systems prepared by mechanical mixing of carbonate with ferric oxide or potassium carbonate, ferric oxide and cerous carbonate of different concentration, as well as catalyst samples obtained by the method of wet mixing of ferric oxide with potassium, magnesium, calcium, molybdenum and cerium salts, were analyzed. Produced mass was molded with hydraulic press into pellets with their subsequent drying-out and annealing on air at given temperatures and time.

For preparation of test samples the following substances were used: commercial potassium carbonate sesquihydrate (first grade), ferric oxide (≥98 wt% of iron in terms of Fe_2O_3), cerous carbonate (47% total oxides of rare earth elements), magnesium carbonate $3MgCO_3 \cdot Mg(OH)_2 \cdot 3H_2O$ (≥24.5% Mg), calcium carbonate (≥98%), ammonium molybdate tetrahydrate (≥98%).

X-ray phase analysis (XPA) of all samples was performed with a DRON-2 type updated diffractometer (CuK_α radiation, graphite monochromator, $2\Theta = 5° - 60°$, step 0.5°, 30 kV, 15 mA, exposure time 3 s). Average coherent scattering region (CSR) was determined based on diffraction lines broadening.

Synchronous thermo analyzer CTA-409 PC Luxx (Netzsch, Germany) was used for thermal analysis. Experiments were carried out on air in temperature range of 25°C to 1100°C by heating a sample with the rate of 10°C/min.

Particles size distribution and specific surface were determined with laser microanalyser of particles Analysette –22 (Fritzsch, Germany). The technique applied in laser particles determinant is based on diffraction pattern analysis principle and allows to estimate particles size distribution from 0.1 μm to 500 μm.

The analysis of model systems and catalysts samples for mono ferrite and potassium carbonate content was performed by the method of selective chemical analysis [10]. Relative error of the method for determination of potassium mono ferrite content and potassium carbonate content is 5% - 7% and 10% - 12% correspondingly.

Activity tests in dehydrogenation reaction of methyl butanes to isoprene were performed in laboratory flow type reactor (volume of catalyst charge is 40 cm^3) in the range of reaction temperatures of 600°C to 615°C at bulk velocity of methyl butenes 1 h^{-1}, under atmospheric pressure, by dilution of raw material with vapor basing on the 1:20 mole ratio C_5H_{10}:H_2O. Contact gas composition and isoamilene fraction were analyzed by means of chromatography using two gas-liquid chromatographs with thermal conductivity detector.

Composition of light gases (H_2, O_2, N_2, CH_4 and CO) was determined in the column filled with NaX molecular sieves, and hydrocarbon mixture and CO_2 composition was determined in the column filled with a 30% n-butyric acid triethylene glycol (NBTEG) diatomite-based sorbent (0.16 - 0.25 mm fractions).

Catalyst activity ($Ac_5\mathbf{H}_8$) was classified by isoprene yield to missed methyl butenes calculated according to the following formula:

$$Ac_5H_8 = \left(Cc_5H_8 \text{(contact gas)} / \left(Cc_5H_8 \text{(starting)} + Cc_5H_8 \text{(starting)}\right)\right) \cdot 100\%$$

Selectivity on isoprene ($Sc_5\mathbf{H}_8$) was characterized on isoprene yield to decomposed isoamylenes calculated according to the formula:

$$Sc_5H_8 = \left(Cc_5H_8 \text{(contact gas)} / \left(\left(Cc_5H_{10} \text{(starting)} + Cc_5H_8 \text{(starting)}\right) - Cc_5H_{10} \text{(starting)}\right)\right) \cdot 100\%.$$

Dehydrogenation products composition measurement accuracy evaluation was made using mathematical statistics methods, developed for small number of definitions (n < 30). Average arithmetic error of five measurements for each component and average squared error of individual measurement were calculated. Relative squared error is ±2%.

RESULTS AND DISCUSSION

At the first stage of our study the model systems consisting of ferric oxide and potassium and/or cerium salts are examined. By thermal analysis of potassium carbonateferric oxide and potassium carbonate-ferric oxide-ceric oxide systems, heat effects and mass losses caused by salts decomposition, phase transitions and interaction of parent compounds were determined. The results obtained are presented in Figure 1 and Table 1.

Endothermal effects in the range of temperatures up to 200°C are determined by removal of water physically adsorbed and crystallized from both potassium and cerium salt crystal hydrates. Maximal mass loss in this region is observed for three-component system. Endothermal effect in the range of 600°C to 800°C found for Fe_2O_3- K_2CO_3 and Fe_2O_3-K_2CO_3-$Ce_2(CO_3)_3$ systems is caused by ferrite formation, reactions (1) and (2) [9], as evidenced by the presence of potassium polyferrite diffracted lines with d = 11.90, 5.95, 2.83 Å etc. in roentgenogram of samples annealed at 800°C.

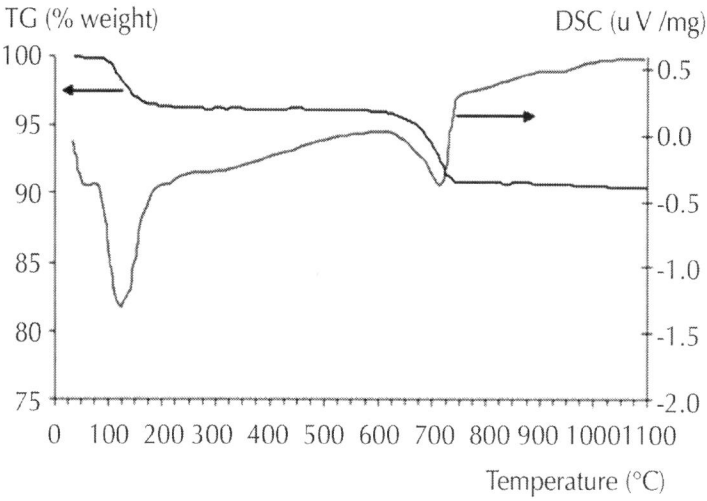

(a)

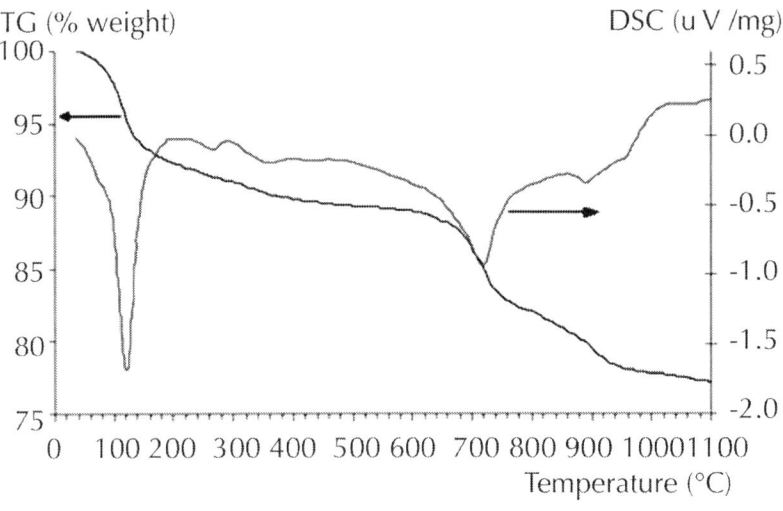

(b)

Figure 1: TG and DSC curves of model mixtures composed of ferric oxide 80% and potassium carbonate 20% (**a**), and ferric oxide 70%, potassium carbonate 20% and ceric carbonate 10% (b).

$$Fe_2O_3 + K_2CO_3 \xrightleftharpoons{500°C - 650°C} K_2Fe_2O_4 + CO_2, \quad (1)$$

$$11Fe_2O_3 + K_2CO_3 \xrightarrow{>650°C} K_2Fe_{22}O_{34} + CO_2. \quad (2)$$

All other endothermal effects observed on DSC curve of three-component mixture are related to the following decomposition of $Ce_2(CO_3)_3 \cdot 5H_2O$ [11]:

$$Ce_2O_3 \cdot 3CO_2 \xrightarrow{300°C} Ce_2O_4 \cdot 2CO_2$$
$$\xrightarrow{380°C} Ce_2O_4 \cdot CO_2 \xrightarrow{480°C} [Ce_2O_4 \cdot CO_2]Ce_2O_4$$
$$\xrightarrow{900°C} 2CeO_2 \quad (3)$$

If cerous carbonate is introduced, the shift of endothermal effect minimum in temperature range of 600°C - 800°C does not practically occur. Maximal mass loss ($\Delta G = 7.0\%$) in this range is observed for three-component system either due to CO_2 removal as a result of a great amount of potassium ferrites formation or due to cerous carbonate decomposition. For confirmation of any version, samples annealed at 800°C were tested by selective chemical analysis to reveal if free potassium carbonate and potassium mono ferrite presents [10]. It has been found that the Fe_2O_3-K_2CO_3 system contains 0.79% of mono ferrite and 0.90% potassium carbonate, and 1.59% and 9.57% correspondingly are contained in the Fe_2O_3- K_2CO_3-$Ce_2(CO_3)_3$ system. Therefore, a noticeable mass loss ($\Delta G = 7.0\%$) in temperature range of 600°C - 800°C for cerium-containing sample is explained by simultaneous processes of potassium mono ferrite formation and cerous carbonate decomposition.

The results obtained by selective chemical analysis indicate that introducing CeO_2 to model mixture promotes formation of a great amount of potassium mono ferrite according to the Equation (1).

For determination of polyferrite phase content in systems the XPA method was used. The values of intensity and diffraction line areas relating to $K_2Fe_xO_y$ and α- Fe_2O_3 (Table 2) show that the number of porassium polyferrites decreases whereas free hematite content increases as ceric oxide content rises. It should be noted that hematite initial particles size decreases as CeO_2 grows from 595 Å to 390 Å. This can be explained by entering of cerium into the crystal lattice of ferric oxide [12], as evidenced by widening of diffraction line with d = 3.67 Å (Table 2), with no solid solution formation. It can be assumed that decreased CSR of hematite is a result of microdeformation due to entering of cerium ions into ferric oxide matrix that is in well conformity with literary data [12]. Most probably, ceric oxide prevents potassium polyferrites formation, with potassium monoferrites being mainly formed during the topochemical reaction of hematite and potassium carbonate that results from selective chemical analysis and published data [7].

Table 1: Derivatographic analysis data of model mixtures

Endoeffects in temperature range								25°C - 1100°C
25°C - 200°C		200°C - 400°C		600°C - 800°C		800°C - 1050°C		
T_{min}, °C	ΔG, %	T_{max}, °C	ΔG, %	T_{max}, °C	ΔG, %	T_{min}, °C	ΔG, %	$\Sigma \Delta G$, %
125	3.6	-	-	715	5.1	946	0.3	9.5
125	7.6	270,350	2.6	720	7.0	892,956	4.4	22.8

Table 2: Average values of crystallites and diffraction line areas according to XPA data of model mixtures

CeO_2 content, %	D_{CSR}, Å			S_{peak}, units			Phases present
	d = 3.67 Å (hematite)	d = 11.90 Å (polyferrite)	d = 3.12 Å (ceric oxide)	d = 3.67 Å (hematite)	d = 11.90 Å (polyferrite)	d = 3.12 Å (ceric oxide)	
0	522	458		41	190		$K_xFe_{22}O_{34}$, α-Fe_2O_3
3.7	486	416	215	51	165	50	
6.3	485	403	162	64	126	94	$K_xFe_{22}O_{34}$, α-Fe_2O_3, CeO_2
7.5	450	353	164	79	58	96	
8.7	390		171	85		134	

Ceric oxide has similar effect on potassium polyferrites. Thus, roentgenogram analysis showed that when ceric oxide concentration grows from 3.7 to 8.7 wt%, the diffracted line of potassium polyferrite 002 moves from d = 11.79 to 12.01 Å. Probably, it might be due to either by introducing **Ce^{4+}**, having greater ion radius compared to Fe^{3+} (0.87 Å and 0.49 Å correspondingly), into $K_2Fe_xO_y$ polyferrite matrix to form point defects or ferrites of different structure $K_2Fe_{10}O_{16}$, $K_2Fe_{14}O_{22}$, $K_2Fe_{22}O_{34}$, and others [5, 13].

Ceric oxide promotes the decrease in CSR of pure hematite (Table 2) and prevents potassium polyferrite formation.

Basing on the aforesaid, ceric oxide influence on phase composition of iron oxide catalysts can be proposed:

$$Fe_2O_3 + K_2CO_3 \xrightarrow{t} \langle KFe_2O_4 \leq K_xFe_{22}O_{34} \rangle$$

$$Fe_2O_3 + CeO_2 \xrightarrow{t} [Fe_2O_3 def]$$

$$\xrightarrow{+K_2CO_3 \cdot t} \langle KFeO_2 \geq K_xFe_{22}O_{34} \rangle$$

Particle size analysis was used to study ceric oxide effect on particles size distribution and on specific surface of ferric oxide. When ferric oxide is annealed up to 950°C, the shift of particles size distribution peak occurs to greater values from 0.5 to 0.85 μm because of recrystallization. As a result of introducing ceric oxide (6.7 wt %), a slight increase in particle sizes of the system Fe_2O_3-CeO_2 is observed (D_{cp} = 0.81 μm) in comparison with pure ferric oxide particle sizes (D_{cp} = 0.66 μm).

Particle size analysis of model mixtures containing potassium carbonate, (Figure 2 and Table 3) demonstrated that the Fe_2O_3-K_2CO_3 system has monomodal particle size distribution with a peak at 245 μm, a low value of specific area and a high bulk weight (0.40 m²/g and 1.65 cm³/g correspondingly) (Table 3). As a result of topochemical reaction of potassium carbonate with ferric oxide in the absence of CeO_2, a primary formation of potassium polyferrites having the average size of crystallites (DCSR = 458 Å), close to hematite average size (BCSR = 522 Å) probably takes place. When interacted (adhered to each other), these potassium polyferrites facilitate particles agglomeration. In turn, this promotes a significant increase of particle sizes in the Fe_2O_3-K_2CO_2 system compared to pure α-Fe_2O_3 annealed under similar conditions (Table 2).

When ceric oxide is added up to 6.3 wt%, the peak of secondary particles size distribution shifts to smaller values in the region of 245 to 25 μm.

If ceric oxide content is higher than 6.3%, no shift occurs, however the growth if its intensity is observed. Figure 2 shows that fine fraction content increases and particles with sizes of more than 200 μm disappear. Here, the distribution becomes bimodal and the presence of two peaks is observed in the range of 1.4 to 25 μm. The average particles diameter decreases from 195 to 17 μm. As a consequence, the bulk density decreases from 1.56 to 1.22 cm^3/g, and specific surface rises from 1.22 to 3.69 m^2/g. The observed effect can be explained by the fact that as CeO$_2$ is introduced the polyferrite phase (predominantly monoferrites) formation occurs to a less extent, that results in greater degree of particles dispersion of Fe$_2$O$_3$-K$_2$CO$_3$-**Ce$_2$(CO$_3$)$_3$** system.

Most probably, introducing ceric oxide into model systems promotes the formation of less crystallized ferric oxide as evidenced by its CSR decreased from 522 to 390 Å, stipulating the formation of predominantly finedispersed potassium monoferrite. In accordance with literary data [7], catalyst activity grows as monoferrite phase content increases. It can be assumed that the increase in ceric oxide concentration in ferric oxide catalyst will affect its performance characteristics.

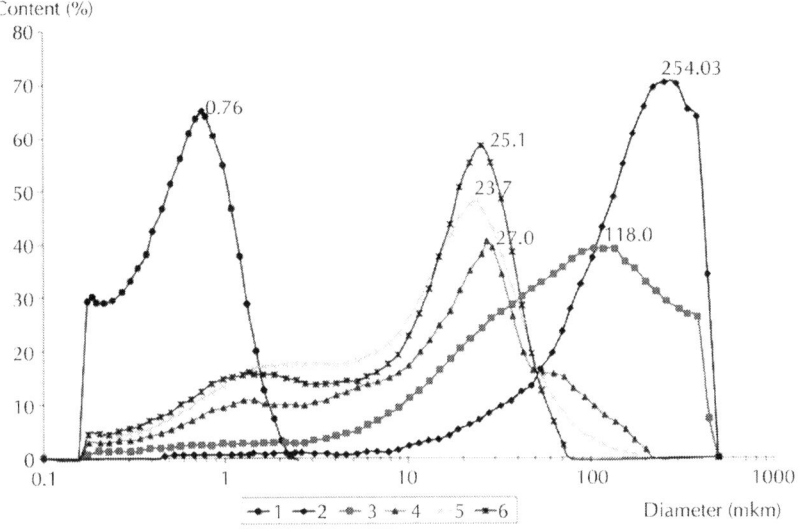

Figure 2: Curves of particle size distribution of ferric oxide (1) and mixtures containing potassium carbonate and **CeO$_2$** in the amount of (2) 0 wt%; (3) 3.7 wt%; (4) 6.3 wt%; (5) 7.5 wt%; (6) 8.7 wt%, annealed at 750°C.

For confirmation of this assumption, catalyst samples with the similar Fe$_2$O$_3$/K$_2$CO$_3$ ratio and different ceric oxide content (from 6.3 to 12 wt %) were prepared.

XPA results obtained for test samples are listed in Table 4. It can be seen that all alterations of phase composition observed on model systems take place also for test catalyst samples. Thus, when ceric oxide content grows the formation of polyferrite phase decreases, the content of both free ferric oxide and potassium monoferrite increases, as evidenced by the values of diffracted line areas of these phases (Table 4) and by the results of selective chemical analysis (Table 5).

Optimum cerium content (8.7 wt %) in catalyst samples with given formulation was found, where catalytic indexes have extreme values (46.3% and 89.8%). The increase in cerium concentration results in hematite content growth according to XPA data (Table 4), that leads to decreased selectivity of catalyst samples from 90% to 87%. As reaction selectivity depends on catalyst porous structure [14], it can be assumed that further increase in ceric oxide content the noticeable sample texture modification occurs.

As Table 5 shows, if ceric oxide concentration increases the activity of catalyst samples rises that can be associated with the increase in potassium monoferrite content. This is evidenced by correlation of $KFeO_2$ concentration with catalyst activity.

CONCLUSIONS

Model systems with different ceric oxide content were analyzed by means of differential thermal analysis, particle size analysis, and X-ray phase analysis. It was shown, that introducing of ceric oxide promotes the decrease in hematite CSR. It was found that in case of Fe_2O_3-K_2CO_2 system the predominant formation of potassium polyferrites is observed, by introducing CeO_2 from 3.7 to 8.7 wt% the content of potassium monoferrites rises along with $K_2Fe_xO_y$ formation, that leads to particle dispersion from D_{cp} = 195.4 **μm** (0 wt% CeO_2) up to 16.8 **μm** (8.7% CeO_2). According to XPA data obtained for catalyst samples, regular relationships typical for model systems are observed. Introducing of ceric oxide leads to increased content of the active phase-potassium monoferrite, stipulating catalytic activity growth. The optimal ceric oxide content of 8.7 wt% was found.

Table 3: Particle size composition, specific surface and bulk weight of model samples

Component	CeO_2 content, wt.%	Particle size range, μm	$D_{average}$, μm	D_{max}, μm	$S_{specific}$, m^2/g	ρ, cm^3/g
α-Fe_2O_3	0	0.1-50	0.9	0.8	8.3	1.28
α-Fe_2O_3-K_2CO_3	0	0.1-500	195.4	245.0	0.40	1.65
α-Fe_2O_3-K_2CO_3-$Ce_2(CO_3)_3$	3.7	0.1-500	108.0	110.0	1.22	1.56
α-Fe_2O_3-K_2CO_3-$Ce_2(CO_3)_3$	6.3	0.1-500	101.0	27.0	2.67	1.50
α-Fe_2O_3-K_2CO_3-$Ce_2(CO_3)_3$	7.5	0.1-200	18.4	24.0	3.69	1.48
α-Fe_2O_3-K_2CO_3-$Ce_2(CO_3)_3$	8.7	0.1-100	16.8	25.0	3.26	1.22

Table 4: Average values of crystallites and diffraction lines area according to XPA data of catalyst samples with various ceric oxide content annealed at 750°C during 3 hours

CeO_2 content, %	D_{CSR}, Å		S_{peak}, units				$S_{K2FexOy}/S_{Fe2O3}$	Phases present
	d = 3.67 Å (hematite)	d = 11.90 Å (poly ferrite)	d = 3.12 Å	d = 3.67 Å (hematite)	d = 11.90 Å (poly ferrite)	d = -3.12 Å (ceric oxide)		
0	428	283		21.3	98.6		4.6	$K_xFe_{22}O_{34}$, α-Fe_2O_3
6.3	486	307	161	14.2	89.7	74.7	6.3	
7.5	477	305	158	26.7	88.8	107.1	3.3	$K_xFe_{22}O_{34}$, α-Fe_2O_3, CeO_2
8.7	459	269	187	38.7	41.0	128.9	1.1	
12.0	400	245	203	62.3	11.1	163.5	0.2	

Table 5: Catalytic properties of catalyst samples with various ceric oxide content as a function of potassium monoferrite concentration

CeO_2 content, %	Catalytic properties, %		$KFeO_2$ concentration, %
	Activity	Selectivity	
0	32.4	88.0	0.81
6.3	44.4	90.2	1.32

7.5	44.0	89.9	1.38
8.7	46.3	89.8	1.56
12.0	47.0	87.2	1.60

ACKNOWLEDGEMENTS

The authors acknowledge funding support from the Ministry of Education and Science of Russian Federation.

REFERENCES

1. N. V. Dvoretsky, E. G. Stepanov, G. R. Kotelnikova and V. V. Yun, "Formation Catalystically Active Ferrite of Potassium," Chemical Fundamentals of Catalysts Formation: Interuniversity Collected Scientific Papers, Kinetics and Catalysis Affairs, Ivanovo, 1988.
2. T. Hirano, "Active Phase in Potassium-Promoted Iron Oxide Catalyst for Dehydrogenation of Ethylbenzene," Applied Catalysis, Vol. 26, 1986, pp. 81-90. doi:10.1016/S0166-9834(00)82543-9
3. I. Serafin, A. Kotarba, M. Grzywa and Z. Sojka, "Quenching of Potassium Loss from Styrene Catalyst: Effect of Cr Doping on Stabilization of the $K_2Fe_{22}O_{34}$ Active Phase," Journal of Catalysis, Vol. 239, No. 1, 2006, pp. 137-144. doi:10.1016/j.jcat.2006.01.026
4. V. M. Busygin, H. H. Gilmanov and S. V. Trifonov, "Catalyst for Dehydrogenation of Olefin and Alkylaromatic Hydrocarbons," RF Patent No. 226675, 2004.
5. M. Baier, O. Hofstadt, W. J. Pöpel and H. Petersen, "Catalyst for Dehydrogenating Ethylbenzene to Produce Styrene," US Patent No. 6551958, 2003.
6. N. Dulamita and A. Maicaneanu, "Ethylbenzene Dehydrogenation on Fe_2O_3-Cr_2O_3-K_2CO_3 Catalysts Promoted with Transitional Metal Oxides," Applied Catalysis A: General, Vol. 287, No. 1, 2005, pp. 9-18. doi:10.1016/j.apcata.2005.02.037
7. E. G. Stepanov, "Scientific Basics of Disintegrational Technology of Manufacturing Fresh Catalysts and Processing of Deactivated

Catalysts of Petrochemical Processes," Dr. Thesis, Yaroslavl State Technical University, Yaroslavl, 2005.

8. A. Trovarelli, C. Leitenburg, M. Boaro and G. Dolcetti, "The Utilization of Ceria in Industrial Catalysis," Catalysis Today, Vol. 50, No. 2, 1999, pp. 353-367. doi:10.1016/S0920-5861(98)00515-X

9. J. Surman, D. Majda, A. Rafalska-Lasocha, P. Kustrowski, L. Chmielarz and R. Dziembaj, "Potassium Ferrites Formation in Promoted Hematite Catalysts for Dehydrogenation. Thermal and Structural Analyses," Journal of Thermal Analysis and Calorimetry, Vol. 65, No. 2, 2001, pp. 445-450. doi:10.1023/A:1017920802391

10. T. G. Zhdanova, R. A. Kuznetsova, A. S. Okuneva and N. K. Loginova, "Separate Identification of Potassium Compaunds in Iron-Chrompotassium Catalyst," Promyshlennost SK (Synthetic Rubber Industry), No. 3, 1986, pp. 5-8.

11. A. I. Leonov, "High Temperature Chemistry of Cerium Oxygen Compounds," Leningrad, 1969.

12. N. C. Pramanic, T. I. Bhuiyan, M. Nakaniahi, T. Fujii, J. Takada and S. I. Seok, "Synthesis and Characterization of Cerium Substituted Hematite by Sol-Gel Method," Materials Letters, Vol. 59, No. 28, 2005, pp. 3783-3787. doi:10.1016/j.matlet.2005.06.056

13. H. H. Gilmanov, A. A. Lamberov, E. V. Dementyeva, E. V. Shatochina, A. M. Gubaidullina and A. V. Ivanova, "Influence of Conditions of Heat Treatment on Formation Polyferrite Phases of Iron-Potassium Catalyst of Dehydrogenation," Neorganicheskie Materialy (Inorganic materials), Vol. 44, No. 1, 2008, pp. 9-15.

14. G. K. Boreskov, "Heterogeneous Catalysis," Moscow, 1988.

Chapter 2

Kinetic Aspects of Gold and Silver Recovery in Cementation with Zinc Power and Electrocoagulation Iron Process

Gabriela V. Figueroa Martinez[1], José R. Parga Torres[1], Jesús L. Valenzuela García[2], Guillermo C. Tiburcio Munive[2], and Gregorio González Zamarripa[1]

[1]Department of Metallurgy and Materials Science, Institute of Technology of Saltillo, Saltillo, México

[2]Department of Chemical Engineering and Metallurgy, University of Sonora, Hermosillo, México

ABSTRACT

The Merrill-Crowe or Cementation process is used for concentration and purification of gold and silver from cyanide solutions. Among other available options for recovery of precious metals from cyanide solutions, Electrocoagulation (EC) is a very promising electrochemical technique for the recovery of this precious metals. In this work first, an introduction to the fundamentals of the Merrill Crowe and EC process are given, then Kinetic aspects conditions and results of the both process, for the removal of gold and silver from cyanide solutions are presented, and finally the characterization of the solid products formed during the EC process with X-ray Diffraction and SEM are shown. Results suggests that The cementation of both gold and silver by suspended zinc particles conforms to well-behaved fist order kinetics and for the EC process the results show that is an excellent option to remove Au and Ag from cyanide solution by using iron electrodes. Finally, 99.5% of gold and silver were removed in the experimental EC reactor, and it was achieved in 5 minutes or less.

INTRODUCTION

Cyanide leaching processes have been used by the mining industry for over 150 years in the extraction of noble metals, the popularity of cyanidation is based mostly on the simplicity of the process. Elsner in Germany in 1846 studied the dissolution of gold in cyanide aqueous solution and noted that atmospheric oxygen played an important role during dissolution of gold [1]. Also, sodium cyanide in an alkaline solution is a strong solvent for gold and silver, most mill operators use it to dissolve fine gold particles with a practical maximum size is no greater than 50 microns. In most cyanidation operations, the gold particles require 24 to 72 hr for complete dissolution in slurry or pulp of about 50 percent solids. Extremely large leaching reactors known as Pachuca tank in with the finely ground ore was agitated with the alkaline cyanide leaching agent and equipped with compressed air injection in the pulp had been designed to dissolve the gold and silver. It is widely accepted that the gold cyanidation process can be represented by the classic Elsner's Equation (1). The mechanism and kinetics have been discussed in several papers and reviews [1,2].

$$4Au + 8CN^- + O_2 + 2H_2O = 4Au(CN)_2 + 4OH^- \tag{1}$$

Silver, similarly, dissolves readily in dilute cyanide solutions in the presence of oxygen. However, since silver in Mexicans ores occurs as argentite the cyanidation and sulphide oxidation reaction are as follows:

$$Ag_2S + 4CN^- = 2Ag(CN)_2^- + S^{2-} \qquad (2)$$

$$S^{2-} + CN^- + \frac{1}{2}O_2 + H_2O = CNS^- + 2OH^- \qquad (3)$$

The cyanide concentration determines the rate of anodic gold dissolution while the oxygen reduction rate is dependent on the concentration of dissolved oxygen. In the cyanidation process, free cyanide ions in solution can be provided only at pH > 9.3. The pH of the pulp can be increased with the additions of alkali hydroxides (NaOH, KOH, Ca(OH)$_2$, etc.), known as proactive alkalis. Also, details of this electrochemical reaction have received considerable attention and under certain circumstances the reaction is limited by the coupled diffusion of CN$^-$ and O$_2$ to the gold surface.

Lixiviation of undesirable base metals, such as copper, iron and arsenic, reduces the efficiency of the process by consuming additional reagents, and necessitates further processing to remove these contaminants [3].

Actually, the two conventional processes for gold and silver recovery from cyanide leach solution are: the Merrill Crowe zinc dust cementation and the carbon adsorption process. In the first process, the efficiency of the reaction is significantly improved by removing dissolved oxygen from the system prior to zinc addition, using a vacuum deaeration technique. This reaction of cementation is sensitive to suspend solids in the leach solution and thus requires solution clarification before cementation. In the activated carbon process, the precious metals are absorbed onto granules of carbon and after loading, they are then stripped of the loaded gold by a hot causticcyanide solution. Then, the cathodes from the carbon adsorption process or the precipitates from the Merrill Crowe process (metal displacements) are then melted in crucible furnaces along with fluxing materials such as borax, niter and silica. The resultant product from smelting is Dore bullion of precious metals typically analyzing more than 95 percent of precious metals.

Each recovery method has advantages and disadvantages. Process selection depends on the specific conditions for particular operation and the facilities already available. Another new process is the Electrocoagulation (EC) technology and has recently been reviewed as an alternative for gold and silver recovery from alkaline cyanide solutions [3]. The EC process is a very promising technique for the recovery of precious metals such as silver and gold: EC needs no chemical reagents, does not generate toxic materials requiring special disposal and this also make it an ecologically viable technique. Literature reviewing showed that the potential of EC as an alternative to traditional treatment recovery of precious metals (silver and gold cyanide) has not yet been exploited.

The Merrill Crowe Process or Cementation Technology

The term cementation comes from a Spanish word meaning "precipitation". This term was first used in 1500 to describe a process for the recovery of copper from aqueous solutions at Rio Tinto in Spain. However, cementation of a noble metal from solution by means of more base, or electronegative, metal has been practiced since ancient times [4]. Near the beginning of the fourth century, Zosimos stated that, when iron is immersed in a solution of a copper salt, it acquires a coating of copper. In the sixteenth century, Paracelsus precipitated silver from silver nitrate by inserting a plate of copper in the solution, and he noted that the copper dissolved. Bergman, in this "De Precipitates Metallics", in the mideighteenth century, observed that the metals precipitate one on another after a certain order of nobility: zinc, iron, lead, tin, copper, silver and gold.

Then, cementation reactions, variously know as metaldisplacement reactions or contact-reduction reactions, are processes in which a metal ion in a solution is reduced to the elemental metallic state with the concurrent oxidation of a more electronegative metal placed into the same solution. This process is one of the most ancient, yet economical and efficient, hydrometallurgical processes known and has been used for the recovery of dissolved metal values from leach solutions, as well as for the purification of electrolytes. Cementation has been extensively applied for the recovery of gold and silver from cyanide solutions, typically followed by the production of Dore metal, which is known as the Merrill-Crowe process. Although cementation of precious metals is widely used in the metallurgy industry, most of the articles published on cementation of gold and silver until about 1980 dealt only with plant practice. Scientific detailed laboratory investigations of the cementation reaction have only been reported during the past three decades. Studies of gold and silver have mostly been confined to the kinetics of gold and silver cementation on rotating discs or cylinders [5].

In spite of this research effort, the mechanistic details of the reactions taking place during cementation and the role of various species in solution in promoting or inhibiting the recovery of precious metals have received little attention and are still somewhat unclear. For example, it is generally accepted that cementation is sensitive both to the alkalinity and to the free cyanide concentration in solution. However, suitable experimental data are not available to describe plant practice in any detail. Only a few fundamental studies have been carried out in the recent past and also, until now, no detailed research has been reported regarding the morphological influence of the deposit structure on the cementation rate.

Review of Cementation Phenomena

In cementation reactions the metal to be reduced from aqueous solutions is more noble, having a greater electron affinity, than the precipitating metal. For example, the recovery or metallic silver by cementation with zinc dust is an electrochemical process involving the oxidation of zinc and the reduction of the silver cyanidation. The overall stoichiometry for the reaction in the case of silver is as follows:

$$2Ag(CN)_2^- + Zn = 2Ag + Zn(CN)_4^- \qquad (4)$$

A similar reaction can be written for gold precipitation. Silver is deposited at cathodic surface sites while zinc dissolves from anodic sites, and electrons are conducted between the two metallic phases as shown in Figure 1.

Cementations Kinetics

Considerable research has been reported on the characteristic features and mechanistic details of cementation reactions and this literature has been reviewed in terms of electrochemical theory and transport phenomena on several occasions [6].

In most cases, the cementation reactions follow firstorder reaction kinetics and generally are limited by diffusion of the noble metal ion through the mass-transfer boundary layer. In some cases the structure and morphology of the reaction product can have a significant effect on passivating the surface in other cases.

Most of the cementation reactions are found to be firstorder diffusion processes [7] with respect to the noble metal ion, and the reaction velocity constant, k, for such a reaction may be computed from the general first-order rate equation:

$$\frac{dc}{dt} = \frac{-kAC}{V} \tag{5}$$

If k is not concentration dependent and the area is unchanging, Equation (5) may be integrated resulting in the integrated first-order expression,

$$\log \frac{C}{C_o} = -\frac{kAt}{2.3V} \tag{6}$$

where C and C_o are the noble metal concentrations at tine t and the initial concentration at time t = 0, respectively;

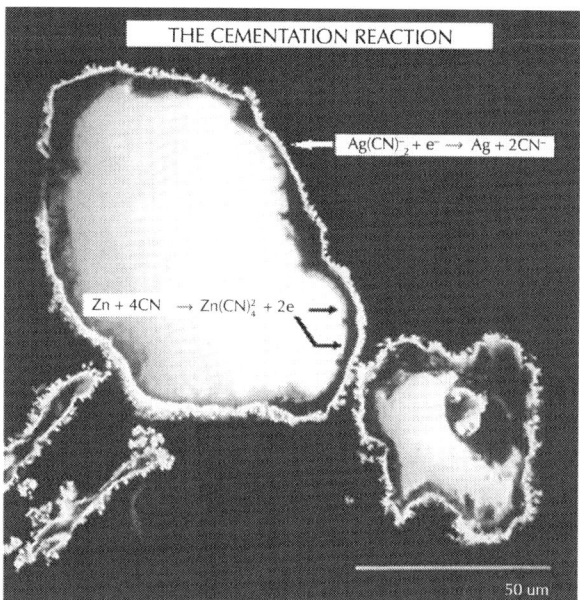

Figure 1: Electrochemical reaction of cementation of silver on zinc particle.

k is the reaction velocity constant (cm/sec); A is reaction surface area (cm^2); and V is the solution volume rate (cm^3). In Equation (5), it can be seen that the cementation rate is a function of the reaction area A. It has been shown by many investigators that in almost all cementation systems the initial exposed geometric area of the precipitant metal can be used in the analysis of initial rate data. However, in a more general sense, the area term is not always that simple to evaluate due to the changing nature of the noble metal deposit which grows on the active metal during the course of the reaction.

Experimental investigations [7] indicate that most cementation reactions are controlled by a mass transfer process, film diffusion, as indicated by the results presented in Table 1.

Notice in Table 1, that for almost all the cementation systems the apparent activation energy is in the range 2 - 6 kcal/mole which would suggest that the cementation reactions are limited by mass transfer in the aqueous phase, with some exceptions (Pb2+/Fe and Pb2+/Cu). Also the reaction velocity constants are of the order 10–2 cm/sec which also supports the position that these cementation reactions are mass transfer controlled.

The Electrocoagulation Technology

Electrocoagulation (EC) has been known as an electrochemical phenomenon since the last century. It has been employed previously for treating many types of waste- water with varying degrees of success [8-10]. The EC process can be considered as an acelerated corrosion process was green rust (GR) is an intermediate product that is responsible for the removal of contaminants (suspended and dissolved solids, metals, organic compounds, etc). EC mechanisms may involve oxidation, reduction, decomposition, deposition, coagulation, absorption, adsorption, precipitation and flotation. EC operates on the principle that hidrolized cations produced electrolytically from iron and/or aluminum anodes enhance the coagulation of contaminants from an aqueous medium.

Table 1: Data for selected cementation aqueous systems at 25°C [7]

System	ΔE_o, V	Activation Energy, Kcal/mole	Reaction Velocity Constant, cm/sec
Ag^+/Cu	0.46	2.0 - 5.0	$2.5 - 6.0 \times 10^{-2}$
$Ag^+/Cu\ (CN^-)$	1.83	3.7 - 5.8	1.5×10^{-2}
$Ag^+/Fe\ (Cl^-)$	1.29	3.0	2.2×10^{-2}
$Ag^+/Zn\ (CN^-)$	0.95	5.5	5.5×10^{-2}
Ag^+/Zn	1.56	2.0 - 6.0	$2.6 - 5.2 \times 10^{-2}$
As^{3+}/Cd	0.65	3.1 ± 1.3	7×10^{-3}
$Au^+/Zn(CN^-)$	0.61	3.1	1.7×10^{-2}
Bi^{3+}/Fe	0.76	4.5 - 7.6	2.9×10^{-2}
Cd^{2+}/Zn	0.36	4.0 - 4.7	$0.54 - 1.1 \times 10^{-2}$
Cu^{2+}/Fe	0.75	3.1 - 5.1	$0.6 - 0.9 \times 10^{-2}$

Cu2+/In	0.83	2.3	5.9×10^{-2}
Cu2+/Ni	0.57	2.7 - 3.7 (14.2 - 19.0)	$0.25 - 1.0 \times 10^{-2}$
Cu2+/Zn	1.10	3.1	$1.6 - 2.1 \times 10^{-2}$
Ni2+/Fe	0.21	7.0	1.4×10^{-4}
Pb2+/Fe	0.31	12.0	-
Pb2+/Zn	0.64	-	0.64×10^{-2}
Pb2+/Cu	0.49	9.5 - 7.4	$0.36 - 2.3 \times 10^{-2}$

The sacrificial metal anodes are used to continuously produce polyvalent metal cations in the vicinity of the anode. These cations facilitate coagulation by neutralizing the negatively charged particles that are carried toward the anodes by electrophoretic motion. Generally in the EC process bipolar electrodes are used [9-11]. It has been reported that cells with bipolar electrodes connected in series operating at relatively low current densities produced iron or aluminum coagulant more effectively. In the EC technique, the production of polyvalent cations from the oxidation of the sacrificial anodes (Al or Fe) and the production of electrolysis gases (O_2 and/or H_2) are directly proportional to the amount of current applied (Faraday's law). The electrolysis gases enhance the flotation of the coagulant material. A schematic representation of the EC process, using iron electrodes, is shown in Figure 2. As mentioned above, the gas bubbles produced by electrolysis carry the gold and silver along with the sludge to the top of the solution where it is collected and removed.

However, it is the reactions of the metal ions that enhance the formation of the coagulant. The metal cations are hydrolyzed, releasing hydrogen ions that result in hydrogen evolution at the cathode, to yield both soluble and insoluble hydroxides that will react with or adsorb gold and silver from the cyanide solution and also contribute to coagulation by neutralizing the negatively charged colloidal particles that may be present at neutral or alkaline pH.

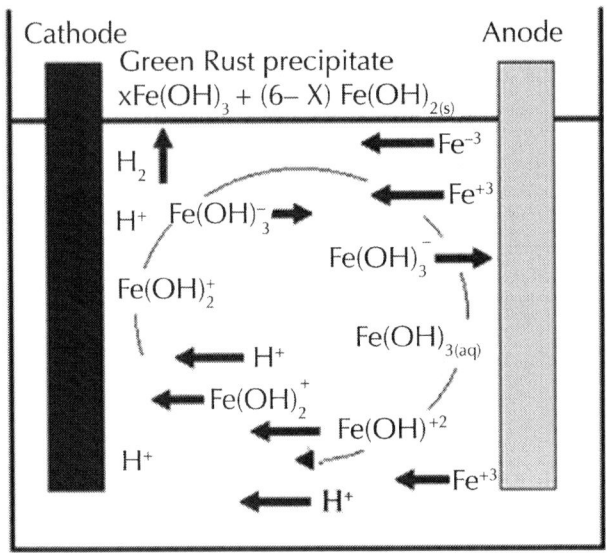

Figure 2: An illustration of the EC mechanism (arrow indicate the migration of ions, the H_2 evolution and the formation of green rust).

This enables the particles to approach closely and agglomerate under the influence of Van der Waals attractive forces. The chemical reactions that have been proposed to describe EC mechanism [12] when using iron electrodes are:

$$Fe \leftrightarrow Fe^{+3} + 3e^- \tag{7}$$

$$Fe(OH)^{+2} + H_2O \rightarrow Fe(OH)_2^{+1} + H^+ \tag{8}$$

$$2H^+ + 2e^- \rightarrow H_{2(g)} \uparrow \tag{9}$$

$$Fe(OH)_2^{+1} + e^- \rightarrow Fe(OH)_{2(aq)} \tag{10}$$

$$Fe(OH)_{2(aq)} + H_2O \rightarrow Fe(OH)_3^{-1} + H^+ \tag{11}$$

$$Fe(OH)_3^{-1} \rightarrow Fe(OH)_{3(aq)} + e^- \tag{12}$$

Overall reaction

$$6Fe + (12 + x)H_2O \rightarrow \frac{1}{2}(12 - x)H_{2(g)} \uparrow \\ + xFe(OH)_3 \times (6 - x)Fe(OH)_{2(s)} \tag{13}$$

The pH of the medium usually rises as a result of this electrochemical process and the Green Rust formed [xFe(OH)$_3$ × (6 − x)Fe(OH)$_{2(s)}$] remains in the aqueous stream as a gelatinous suspension, which can remove the gold and silver from pregnant cyanide rich solutions, either by complexation or by electrostatic attraction of magnetic nanoparticles followed by coagulation and flotation. Generally, in the EC process, bipolar electrodes are used [8]. It has been reported that cells with bipolar electrodes, connected in series operating at relatively low current densities, produce iron or aluminum coagulant more effectively, more rapidly and more economically when compared to chemical coagulation.

EXPERIMENTAL PROCEDURES

Suspended Zinc Particle Experiments

The bath reactor experiments were done in four-necked, one liter glass reaction cells supported in a constant temperature bath. A condenser, stirrer, nitrogen dispersion tube, and sampling derive were placed into a reactor through the openings in the lid. The Teflon stirrer was attached through the center port by means of a Chesapeak stirrer connection. In all studies purified nitrogen was passed through the solution, via a dispersion tube, before and during the experiment to maintain an oxygenfree environment. Tyler sieves were used to prepare monosize fractions of zinc in the range from 70 to 400 mesh.

The specific surface area of the zinc dust was measured using two techniques, the BET technique (Quantasorb Model 0S10, made by Quantachrome Corporation), surface area 255 cm^2/gr, and the air permametry technique (Permaran made by Outokumpu), surface area 216 cm^2/gr. The stirring speed was adjusted manually (1000 RPM) and was checked with a stroboscope. The particulate sample of zinc was washed with acetone and cleaned with acid to remove the surface oxide film before being introduced into the reaction flash. For most of the experiments, one gram of 270 × 400 mesh (d_{50} = 45 μm) particles of zinc were introduced through one of the cover ports, and the reaction was initiated. In order to follow the course of cementation reaction, solution aliquots of 2 ml were taken periodically and analyzed for silver and gold with a DCP (Direct-Current Plasma) Spectrophotometer manufactured by Beckman Instruments.

Experimental Electrocoagulation Details

The EC experiments were performed in a 400 ml beaker size reactor equipped with two carbon steel electrodes (6 cm × 3 cm) that were 5 mm apart. As a source of current and voltage a universal AC/DC adaptor was used. pH was measured with a VWR scientific 8005 pH meter. Gold and silver adsorption onto iron species was investigated with pregnant cyanide solutions provided by Bacis S. A. de C. V. mining group (13.25 ppm Au and 1357 ppm Ag and pH of 8). Analyses were performed by ICP/ Atomic Emission Spectrometry. The conductivity of pregnant solutions was adjusted by adding one gram of NaCl per liter (Fisher, 99.8% A. C. S. Certified, lot #995007).

To identify and characterize the iron species in the solid products, formed during the EC process for the removal of gold and silver using iron electrodes, X-ray diffraction (XRD) and Scanning Electron Microscope (SEM/EDX) were used. Analysis of Au and Ag were conducted to the Bacis solution, by AES. EC was run at 15 Volts (DC) and the corresponding current was of 0.1 Å. EC was run for five minutes, and a sample was taken every minute in order to determinate the removal efficiency for Au and Ag. Solutions and solids from the EC process were separated by filtration through cellulose filter paper. The sludge from the EC was dried either in an oven or under vacuum at room temperature and characterized.

RESULTS AND DISCUSSION

Zinc Dust Particle Size Results

As was shown by Equation (5), the rate of cementation reaction is expected to be a function of the surface area and the concentration of the reacting noble metal ion (gold or silver in the present case). To examine the traditional first order rate expression, the cementation reaction kinetics were studied using monosize zinc particles. Two experiments were conducted at 10 ppm (5×10^{-5} M) gold and 100 ppm (9×10^{-4}) silver in order to study the reaction kinetics at these concentrations. The equilibrium concentration profiles (on logarithmic scale), the results were found to conform to the first-order rate process for over 99% removal as shown in Figure 3 (Au/Zn(CN)– system) and (Ag/Zn(CN)– system). From these data it is evident that the cementation of both gold and silver by zinc conforms to well-behaved fist order kinetics. Also, from these figure, the micrograph of reacted zinc particles can be seen. This micrograph reveals the nature of the silver and gold deposits. The effect of the gold and silver deposits on the rate of cementation reaction product showed that the gold and silver formed uniforms, apparently porous layer around the zinc particle. Table 2 lists the values of the reaction velocity constant for the cementation of gold (10 ppm) and silver (100 ppm) on zinc dust.

Electrocoagulation Results

Running the EC process for gold and silver removal on iron electrodes, gave the results shown in Table 3 and in Figure 4, shows the equilibrium concentration profile of gold and silver removal efficiency and final pHs values vs Time. The obtained results show that EC is an excellent option to remove Au and Ag from cyanide pregnant solution by using iron electrodes. Also, under these conditions, the results show that, when residence time increases from 1 to 3 minutes the silver and gold removal increase to 92% and 99% respectively, this variation occurs at pH values from 9 to 10.7, approximately; these values coincide with the production of magnetic iron, Fe_3O_4 (green rust). Also, it is apparent that the uptake percents of gold/silver increased by increasing the pH value.

The mechanism of the EC process for gold and silver recovery is then physical and chemical adsorption, caused by the highly reactive species and adsorption properties of green rust. These processes do not alter their chemical composition.Figure 4, also shows graphically the pH during the EC the process for gold and silver removal, the pH increment in the solution from 8 to a pH final of 11 is attributed to the hydrogen evolution at the cathode which is then accompanied by alkalinization of the aqueous solution.

Table 2: Comparison of experimental reaction velocity constants with calculated mass-transfer coefficients for suspended particles. NaCN = $10^{-2\,M}$; pH = 10.5; Stirring speed = 1000 rpm; Au = 10 ppm ($5 \times 10^{-5\,M}$); Ag = 100 ppm ($9 \times 10^{-4\,M}$)

System	Particle size, mesh	Experimental reaction velocity constant, cm/sec	Calculated mass-transfer coefficient
Au/Zn(CN)–	400 × 270	1.9×10^{-2}	2.35×10^{-2}
Ag/Zn(CN)–	400 × 270	1.62×10^{-2}	3.2×10^{-2}

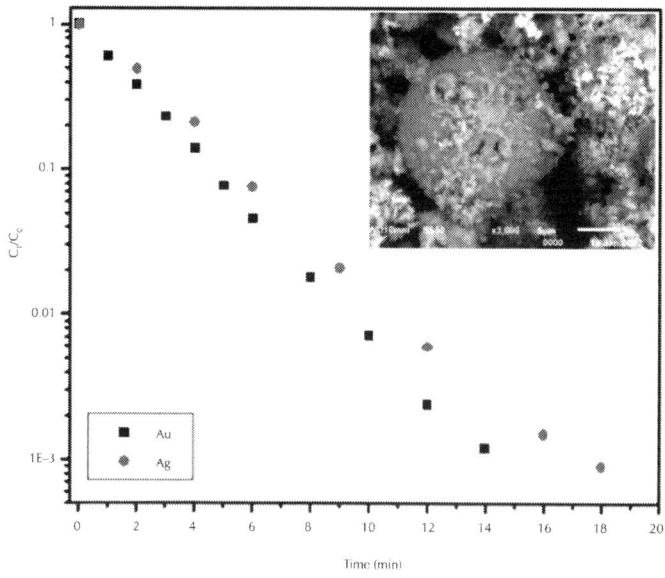

Figure 3: First-order plots showing removal of gold and silver from cyanide solution.

Table 3: Recovery of gold and silver by EC

EC residence time (min)	Au (mg·L–1)	Recovery (%)	Ag (mg·L–1)	Recovery (%)	pH
0	13.25	0	1357.0	0	8.0
1	12.50	5.66	940.0	30.72	9.2
2	10.50	20.37	219.5	83.82	9.5
3	1.00	9.5	9.0	99.33	10.7
4	0.50	96.22	7.0	99.48	11.2
5	0.50	99.24	0.9	99.93	11.5

Product Characterization

X-ray Diffraction Analysis. Diffraction patterns of flocs collected from the experiment with gold and silver, (the sample were ground to a fine powder and loaded into a sample holder) were obtained with a diffracted X-PERT Phillips meters equipped with a vertical goniometer, with a range of analysis 2 10° to 70°. The source of X-rays has a copper anode, whose radiation is filtered with a graphite monochromator (= 1.541838 Å) with scan rate of 0.02° and a duration of 10 seconds per count. The X-ray Diffractometer is controlled by a Gatawey 2000 computer, by PC-APD 2.0 with software for Windows.

Figure 5 shows the ray diffraction pattern of the flocs recovered from the sample of gold and silver, respectively 13.25 mg/L and 1357 mg/L, initial pH 8, 5 minutes of treatment, 0.1 amperes and 15 volts. The species identified were magnetite, lepidocrocite, goehtite, silver and copper hexacyanoferrate. Scanning Electron Microscopy (SEM/EDAX). Figure 6 shows SEM images and EDAX of silver adsorbed on iron species. These SEM and EDAX results show that the surfaces of these iron oxide/oxyhydroxide particles were coated with a layer of silver. It is worth clarifying that, given the low concentration of gold it was impossible to locate any nanoparticle of it.

CONCLUSIONS

The cementation of both gold and silver by suspended zinc particles conforms to well-behaved fist order kinetics. The effect of the gold and silver deposits on the rate of cementation reaction product showed that the gold and silver formed uniforms, apparently porous layer around the zinc particle. The experimental reaction velocity constants correspond to the expected magnitude of the limiting mass-transfer coefficients and support the hypothesis that the cementation reactions under these conditions are mass-transfer limited reactions with recoveries of gold and silver of 99.8%. The obtained results show that EC process is an excellent option to remove Au and Ag from cyanide pregnant solution by using iron electrodes. The X-ray Diffraction, Scanning Electronic Microscopy, techniques demonstrate that the formed species are of magnetic type, like lepidocrocite and magnetite, and amorphous iron oxyhydroxide which adsorbed the silver and gold particles on his surface due to the electrostatic attraction between both metals. The 99.5% of gold and silver were removed in the experimental EC reactor, and it was achieved in 5 minutes or less with a current efficiency of 99.7%.

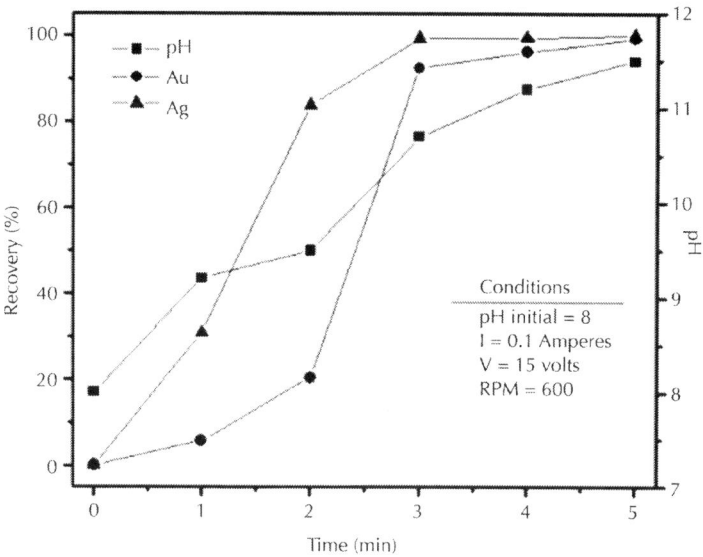

Figure 4: Gold and silver recoveries from Bacis cyanide solutions and pH vs EC residence time.

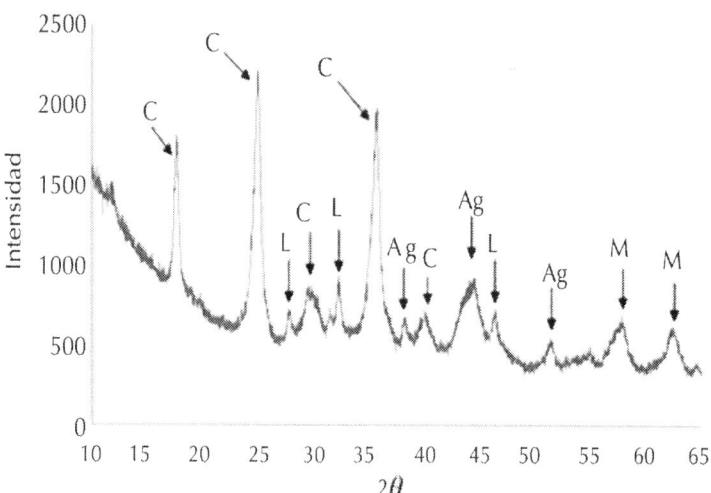

Figure 5: X-ray diffractogram of solids obtained in the recovery of gold and silver C: $Cu_2Fe(CN)6 2H_2O$, A: Silver, M: Magnetite and L: Lepidocrocite.

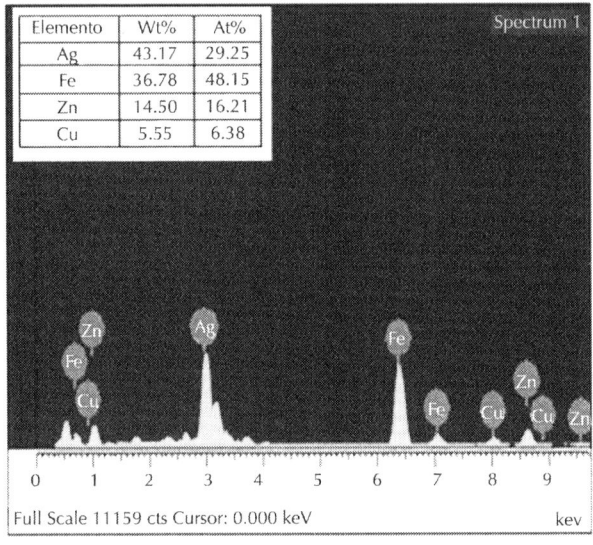

(a)

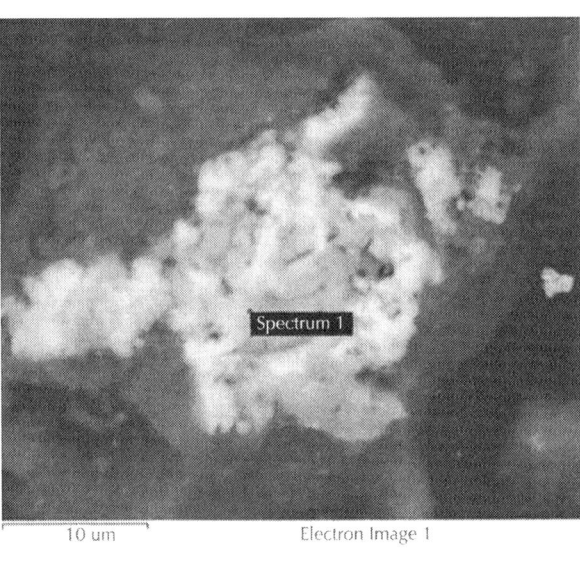

(b)

Figure 6: Chemical composition of solid product as determined by EDX, which shows the presence of silver in the particle of iron.

ACKNOWLEDGEMENTS

The authors wish to acknowledge support of this project to the National Council of Science and Technology (CONACYT) and to Dirección General de Educación Superior Tecnológica (DGEST) from Mexico.

REFERENCES

1. F. Habashi, "A Short History of Hydrometallurgy," Hydrometallurgy, Vol. 79, No. 1-2, 2005, pp. 15-22. doi:10.1016/j.hydromet.2004.01.008
2. J. R. Parga, G. Gonzalez, H. Moreno and J. L. Valenzuela, "Thermodynamic Studies or the Strontium Adsoption on Iron Species Generated by Electrocoagulation," Desalination and Water Treatment, Vol. 37, No. 1-3, 2012, pp. 244-252.
3. J. R. Parga, J. L. Valenzuela, H. Moreno and J. E. Perez, "Copper and Cyanide Recovery in Cyanidation Effuents," Advances in Chemical Engineering and Science, Vol. 1, No. 4, 2011, pp. 191-197. doi:10.4236/aces.2011.14028
4. R. Woods, "Extracting Metals from Sulfide Ores," 2010. http://electrochem.cwru.edu/encycl/art-m02-metals.htm
5. W. Wai, L. Eugene and A. S. Mujumdar, "Gold Extraction and Recovery Processes," 2009. http://www.eng.nus.edu.sg/m3tc/M3TC_Technical_Reports/Gold%20Extraction%20and%20Recovery%20Processes.pdf
6. F. Habashi, "Kinetics and Mechanism of Gold and Silver Dissolution in Cyanide Solution," Montana Bureau of Mines Geological Bulletin, No. 59, 1967, pp. 1-42.
7. S. Gamini, "The Cyanidation of Silver Metal: Review of Kinetics and Reaction Mechanism," Hydrometallurgy, Vol. 81, No. 2, 2006, pp. 75-85.
8. J. R. Parga, D. L. Cocke, J. L. Valenzuela, H. Moreno and M. Weir, "Arsenic Removal via Electrocoagulation from Heavy Metal Contaminated Groundwater in La Comarca Lagunera Mexico," Journal of Hazardous Materials, Vol. 124, No. 1-3, 2005, pp. 247-254. doi:10.1016/j.jhazmat.2005.05.017

9. M. Mollah, P. Morkovsky, J. Gomez, J. R. Parga and D. Cocke, "Fundamentals, Present and Future Perspectives of Electrocoagulation," Journal of Hazardous Materials, Vol. 114, No. 1-3, 2004, pp. 199-210. doi:10.1016/j.jhazmat.2004.08.009

10. M. M. Emamjomeh and M. Sivakumar, "Review of Pollutants Removed by Electrocoagulation and Electrocoagulation/Flotation Processes," Journal of Environmental Management, Vol. 90, No. 5, 2009, pp. 1663-1679. doi:10.1016/j.jenvman.2008.12.011

11. X. Zhao, B. Zhang, H. Liu and J. Qu, "Removal of Arsenite by Simultaneous Electro-Oxidation and ElectroCoagulation Process," Journal of Environmental Management, Vol. 184, No. 1-3, 2010, pp. 472-476. doi:10.1016/j.jhazmat.2010.08.058

12. H. Moreno, D. L. Cocke, J. A. G. Gomes, P. Morkovsky, J. R. Parga, E. Peterson and C. Garcia, "Electrochemical Reactions for Electrocoagulation Using Iron Electrodes," Industrial & Engineering Chemistry Research, Vol. 48, No. 4, 2009, pp. 2275-2282. doi:10.1021/ie8013007

Chapter 3

Gasification Coupled Chemical Looping Combustion of Coal: A Thermodynamic Process Design Study

Sonali A. Borkhade, Preksha A. Shriwas, and Ganesh R. Kale

Chemical Engineering and Process Development Division, National Chemical Laboratory, Pune 411008, India

ABSTRACT

A thermodynamic investigation of gasification coupled chemical looping combustion (CLC) of carbon (coal) is presented in this paper. Both steam and CO_2 are used for gasification within the temperature range of 500–1200°C. Chemical equilibrium model was considered for the gasifier and CLC fuel reactor. The trends in product compositions and energy requirements of the gasifier, fuel reactor, and air reactor were determined. Coal (carbon) gasification using 1.5 mol H_2O and 1.5 mol CO_2 per mole carbon at 1 bar pressure and 650°C delivered maximum energy (−390.157 kJ) from the process. Such detailed thermodynamic studies can be useful to design chemical looping combustion processes using different fuels.

INTRODUCTION

Coal is the most abundantly available cheap fossil fuel worldwide and its reserves are estimated to outlast oil and natural gas reserves [1]. Coal is mainly used for energy generation: in coal fired power plants to produce electricity [2], hydrogen [3–5], and syngas production for FT synthesis or fuel cells [6, 7]. Combustion of coal or coal derived syngas in air results in generation of product gas mixture containing CO_2, N_2, and NO_x [8]. The separation of CO_2 from such gaseous streams is extremely difficult and expensive. Hence these product gases are directly vented to the atmosphere without CO_2 separation. This environmental pollution is a major drawback of energy generation from coal. CO_2 emissions from such processes are mainly responsible for global warming and climate change phenomenon [9]. The 2010 CO_2 emissions have increased to 389.0 ppm and burning of fossil fuels is one of the main causes as reported by the World Meteorological Organization [10]. Such tragic scenarios were foreseen and therefore research in clean energy generation using coal had already started globally. Chemical looping combustion (CLC) technology is a result of such research efforts. CLC uses a solid oxygen carrier (OC) to oxidize the carbon and hydrogen present in the fuel to CO_2 and H_2O, respectively, in an endothermic fuel reactor. Oxides such as NiO, CuO, and Fe_2O_3 and sulphates such as $CaSO_4$ have been widely used as oxygen carriers in chemical looping processes. The reduced OC is regenerated by air oxidation in an exothermic air reactor [11, 12]. Both the reactors are interconnected and operate simultaneously. The energy released in the CLC system is of similar magnitude as direct combustion, but with a crucial advantage of having CO_2 and nitrogen streams completely separated [13, 14]. The pure CO_2 steam can be captured/sequestered easily thereby reducing CO_2 emissions to the atmosphere [15]. Coal mainly contains carbon which is a highly stable solid species and hence it requires high temperature and high energy input for chemical reactions. Further, the coal-OC system is a solid-solid complicated reaction system that has its intrinsic problems like slow reaction kinetics, low reactivity and conversions, need of gaseous medium, and so forth [16]. These problems in coal CLC system prompted researchers to search for better options. A popular option is coal gasification to syngas (CO + H_2) and CLC of the generated syngas. Some research studies have been published with this theme [17–21]. Coal gasification can be done using steam or CO_2 or both [22–26]. Coal gasification using steam (SG) is already a developed technology [27]. The generated syngas is more reactive gaseous species than solid coal. This syngas has high reactivity with OC which can help commercialization of this CLC pathway. CLC of syngas has been successfully studied by researchers with many OCs [28, 29]. NiO is a very popular OC in chemical looping processes and it is used in this study. Coal gasification using CO_2 (dry gasification: DG) results in CO rich gas, while SG provides syngas (H_2+CO). CO_2 is a relatively inert gas, not available in pure form freely, and its reaction with coal is relatively slower enabling complete conversion only at high temperatures. On the other hand, water is cheaply available, but the conversion of water to steam requires huge energy due to high latent heat of water. Hence a right combination of CO_2 and steam might be useful for coal gasification to produce syngas for its further use in CLC. Coal gasification using steam/CO_2 without air is a highly endothermic process. A comprehensive theoretical research study is needed to understand the optimum conditions for the process development involving combined gasification coupled CLC of coal. Such a detailed study has not yet been published in literature. Thermodynamic studies are key starting points for chemical process design [30, 31]. Kinetics of coal gasification rely on the inorganic metal contents of coal, but thermodynamic studies that determine the maximum possible conversions under particular conditions of temperature, pressure, and feed components are not affected by these components/catalysts. Thermodynamic studies based on chemical equilibrium have been done for coal gasification systems [32–36] and also for CLC of coal [37–39]. The systematic thermodynamic study of gasification coupled CLC is presented in this paper to understand the product distribution trends at different (gasifying agent) feed, temperature, and pressure conditions to identify the best quantity of gasification agents leading to process temperatures and pressures for maximizing desired products and minimize the undesired products at maximum extractable energy from the overall process. Such a theoretical study is vital to understand the overall process aspects and reduce the time and efforts on experimentation studies which can further help fast track commercialization.

Process Description

Figure 1 shows the conceptual process design for gasification coupled CLC of coal. The process scheme consists of a gasifier, CLC fuel reactor, and CLC air reactor. Initially, preheated coal, CO_2, and H_2O were fed to the gasification reactor in calculated quantities to produce syngas including minor amounts of CH_4, CO_2, and H_2O. The gasification products were assumed to be in thermodynamic equilibrium at the exit of the gasifier. Complete conversion of coal and maximum syngas production are targeted in the gasifier. The products obtained from the gasifier at high temperature are directly fed to the fuel reactor with calculated quantity of heated OC (NiO) recirculating from the air reactor. It is assumed that the NiO oxidizes CO, H_2, and CH_4 to CO_2 and H_2O in the CLC fuel reactor. The moles of OC (NiO) for fuel reactor are varied depending on the input moles of CO, H_2, and CH_4 from the gasifier. Complete conversion of syngas and CH_4 to CO_2 and H_2O is desired; however the conversion is limited by thermodynamic equilibrium. Coke formation can also take place in the CLC fuel reactor. The OC (NiO) gets reduced (to Ni) in the CLC fuel reactor. The products of the CLC fuel reactor pass through a gas-solid separator and the gaseous products mainly containing CO_2 and H_2O may be partly recycled to the gasifier (if pure) and the rest can be cooled for CO_2 separation and sequestration with heat recovery. The solid products of the fuel reactor mainly containing OC (and coke) are transferred to the air reactor, in which preheated air is added for complete oxidation of coke to CO_2 and regeneration of reduced OC. It is assumed that these oxidation reactions go to complete conversion. The regenerated NiO is separated from the air reactor product stream by a gas-solid separator and is recycled back to the fuel reactor. The air reactor is the major source of heat in the entire process. It is assumed that the three reactors (gasifier, fuel, and air reactor) operate at same temperature and pressure. This study assumes energy for chemical reactions (including heating/cooling) only and other energies such as OC transport energy, coal feeder energy, gas-solid separator energy, and heat losses are not considered in this basic theoretical study.

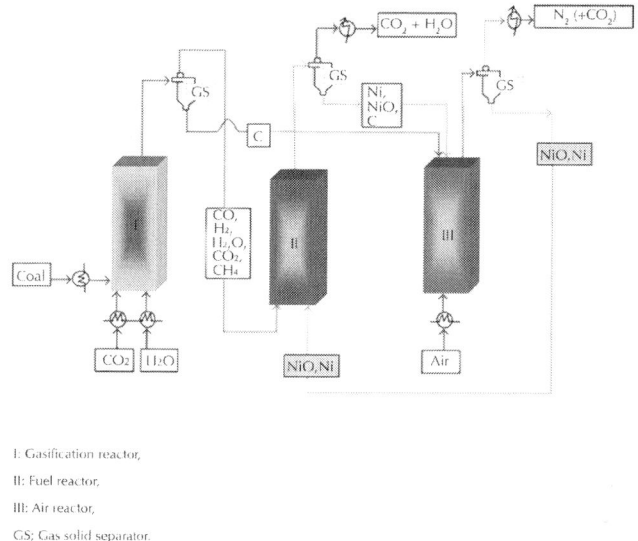

I: Gasification reactor,
II: Fuel reactor,
III: Air reactor,
GS: Gas solid separator.

Figure 1: Process diagram for gasification coupled chemical looping combustion of coal.

Process Methodology

HSC Chemistry software version 5.11 is used to generate the thermodynamic equilibrium data for gasifier and CLC fuel reactor in this process design study [40, 41]. Thermodynamic equilibrium calculations in the Gibbs routine of HSC Chemistry are done using the Gibbs free energy minimization method. The Gibbs program finds the best combination of most stable species where the Gibbs free energy of the system achieves its minimum at a fixed mass balance (a constraint minimization problem), constant pressure, and temperature. Hence chemical reaction equations are not required in the input. The carbon content (in weight) in coal is variable, generally varying from 35–85% depending on its origin (American, Australian, Chinese, etc.). Coal also contains minor amounts of hydrogen (~5%), oxygen (~7%), sulfur (<1%), ash, and so forth. Generally carbon, hydrogen, and sulphur can contribute to energy generation, but the quantity of hydrogen and sulphur is not much compared to the carbon content of coal.

Hence only pure carbon (solid) is used to model coal in this study. This assumption is made to compare the results of this novel process study with standard heat of combustion of carbon which is a unique value compared to the coal combustion values which vary a lot. Species such as C (s), CO_2 (g), H_2 (g), CO (g), H_2O (g), CH_4 (g), H_2O (l), O_2 (g), NiO (s) and Ni (s), which are usually found in the gasification and CLC reaction systems are considered in this study. The input species to the gasifier were C (s), H_2O (g), and CO_2 (g) reacting to give the products. The material balances are done by Equilibrium Composition module of HSC Chemistry and these results were used to calculate reaction enthalpy by Reaction Equation module of HSC Chemistry software. The results may be slightly different using different software as all softwares use their own databank of chemical properties. The results presented are within reasonable error limit (±1%). Any other inert in feed, by-product formation is not considered in this study. The gasification reaction chemistry for coal is well established in chemical literature and hence no such details are presented in this paper. 1 mol carbon has been used as basis for all calculations in the temperature range of 500–1200°C for the entire process. Coal gasification is done using steam and CO_2. The feed gasifying agent-to-carbon ratio (GaCR) ranging from 1 to 3 was selected for the study. Combined gasification of carbon (CG) using both CO_2 and H_2O is also considered in this study and its intermediate steps for increase in CO_2 moles (with simultaneous decrease in steam moles) for constant GaCR ratios were also considered in the calculations. These feed conditions for CO_2 and steam input per mole coal input are shown in Table 1. These inputs were used to calculate the thermodynamic equilibrium compositions in the process reactors and are discussed in detail in the next section with their respective reactor energies. The oxidation reaction in the air reactor is highly exothermic. It is assumed that the air supplied to the air reactor is in stoichiometric amount given by the reactions

$$Ni + 0.5O_2 = NiO \qquad (1)$$

$$C + O_2 = CO_2 \qquad (2)$$

The equilibrium compositions of the coal gasifier and CLC fuel reactor were used to calculate the energy involved in those reactors. For example, the equilibrium product composition for case B2 in the gasifier was [0.51CO$_2$ (g), 0.49H$_2$O (g), 0.99CO (g), 1.01H$_2$ (g)] at 800°C. This product composition was used to formulate the following reaction:

$$C + 1.5H_2O\ (g) + 0.5CO_2\ (g) = 0.51CO_2\ (g) + 0.49H_2O\ (g)$$
$$+ 0.99CO\ (g) + 1.01H_2\ (g) \quad (3)$$

This reaction was fed to the Reaction Equation module of HSC Chemistry software and the reaction enthalpy of $\Delta H = 135.43$ kJ at 800°C was obtained. Similar strategy was followed for CLC fuel reactor material and energy balances. The calculations for the air reactor were done using reactions equations only as complete conversion (no equilibrium) was assumed. The general calculations required to determine energy requirements of heating and cooling of chemical species were done manually using standard data [42].

Table 1: Gasification conditions

Feed Conditions	Input moles of carbon	Input moles of CO$_2$	Input moles of H$_2$O	Gasifying agent to carbon ratio (GaCR) ((H$_2$O + CO$_2$)/C)
A1	1	0	1	1
A2	1	0.25	0.75	1
A3	1	0.5	0.5	1
A4	1	0.75	0.25	1
A5	1	1	0	1
B1	1	0	2	2
B2	1	0.5	1.5	2
B3	1	1	1	2
B4	1	1.5	0.5	2
B5	1	2	0	2
C1	1	0	3	3
C2	1	1	2	3
C3	1	1.5	1.5	3
C4	1	2	1	3
C5	1	3	0	3

RESULTS AND DISCUSSIONS

Effect of Pressure

Pressure is an important process parameter. The effect of pressure on gasification coupled CLC of coal was investigated for 1, 5, and 10 bars. Initially, the effect of pressure on carbon conversion and syngas yield in the gasifier was studied. It was observed that the theoretical carbon conversion reached a maximum value (approximately 100%) at higher temperatures (>600°C). Hence the effect of pressure was studied at 600°C only. Figure 2(a) shows the carbon conversion at 600°C for the 15 feed input conditions (A1–C5). It was observed that the carbon conversion generally decreased with increase in pressure for all cases. It was also observed that the carbon conversion in CG at constant pressure generally decreased with increase in feed CO_2 moles at constant GaCR for all pressures. It was also seen that the carbon conversion increased as the feed steam to carbon ratio (SCR) increased from 1 to 3 for all pressures for SG cases (A1, B1) and sometimes saturated at 100% (B1, C1). Similar observation was also noted for DG cases (A5, B5, and C5) with increase in the feed CO_2 to carbon ratio (CCR) from 1 to 3. Figure 2(b) shows the syngas yield for the different feed conditions at 600°C. It was observed that the syngas yield decreased with increase in pressure for all cases. It was also observed that the syngas yield in CG at constant pressure decreased with increase in feed CO_2 moles at constant GaCR for all pressures. It was also seen that the syngas yield increased as the respective feed GaCR increased from 1 to 3 for all pressures for SG cases (A1, B1, C1) and for DG cases (A5, B5, C5).

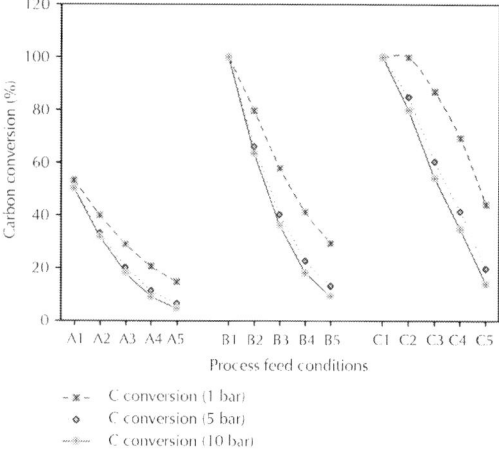

(a)

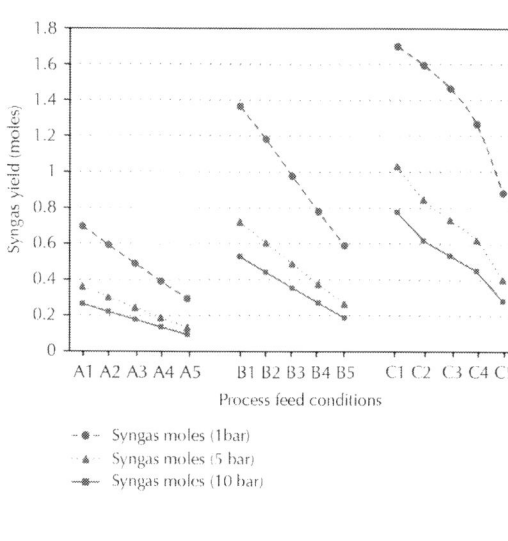

(b)

Figure 2: (a) Carbon conversion in gasifier at 600°C. (b) Syngas yield in gasifier at 600°C.

Considering the negative effect of pressure on carbon conversion and syngas yield in the gasifier, it was noted that the maximum carbon conversion with maximum syngas yield can be achieved at low pressure (1 bar). Hence 1 bar pressure was selected for further analysis of gasification coupled CLC process. The carbon conversion in the gasifier for the different feed conditions at 1 bar pressure is shown in Table 2. It was observed that the carbon conversion reached its maximum (100%) as the gasification temperature increased from 500–1200°C. It was generally observed that higher GaCR required relatively lower temperatures for 100% carbon conversion and the 100% carbon conversion in SG occurred at relatively lower temperatures than analogous DG cases. The minimum temperatures for 100% carbon conversion (HCCT: Hundred percent Carbon Conversion Temperature) were important and sufficient for this process design. Hence these HCCTs (Bold in Table 2) were selected for further process analysis for the respective feed conditions. Thus the pressure and temperature for the process reactors were fixed for further process design. The product/energy analysis of the individual reactors depended solely on the feed composition conditions and is discussed in the next sections.

Table 2: Carbon conversion in gasifier

Temperature (°C) Feed Conditions	500	550	600	650	700	750	800	850	900	950	1000	1050	1100	1150	1200
		Carbon conversion %													
A1	46	48	54	62	73	84	92	96	98	99	100	100	100	100	100
A2	29	33	40	51	65	79	88	94	97	98	99	100	100	100	100
A3	16	21	29	41	57	73	85	93	96	98	99	99	100	100	100
A4	8	13	21	34	51	69	83	91	96	98	99	99	100	100	100
A5	3	7	15	28	46	65	81	90	95	97	99	99	100	100	100
B1	91	97	100	100	100	100	100	100	100	100	100	100	100	100	100
B2	57	66	80	100	100	100	100	100	100	100	100	100	100	100	100
B3	32	42	58	83	100	100	100	100	100	100	100	100	100	100	100
B4	16	25	42	67	100	100	100	100	100	100	100	100	100	100	100
B5	7	15	30	55	91	100	100	100	100	100	100	100	100	100	100
C1	100	100	100	100	100	100	100	100	100	100	100	100	100	100	100
C2	72	86	100	100	100	100	100	100	100	100	100	100	100	100	100
C3	49	63	88	100	100	100	100	100	100	100	100	100	100	100	100
C4	31	45	70	100	100	100	100	100	100	100	100	100	100	100	100
C5	10	22	45	83	100	100	100	100	100	100	100	100	100	100	100

Products of the Gasifier

The product composition of the gasifier obtained for the various (A1 to C5) feed conditions at 1 bar pressure and respective HCCTs was analyzed and plotted in Figure 3.

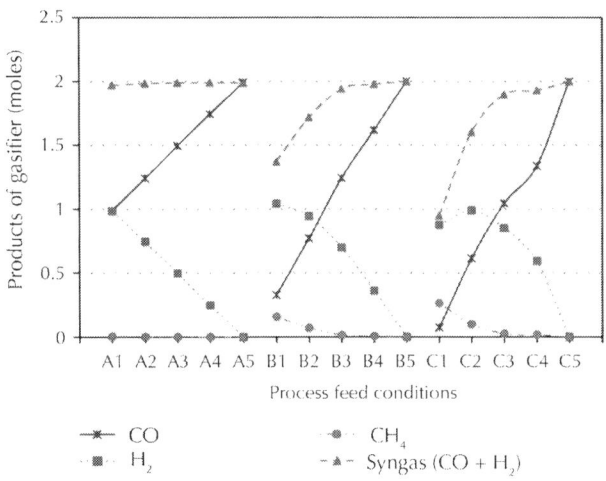

Figure 3: Products of gasifier at different feed conditions.

Syngas

Syngas is the most desired product of the coal gasifier. As seen in Figure 3, the syngas yield in CG generally increased with increase in feed CO_2 moles at constant GaCR and reached a maximum for almost all cases. The maximum syngas yield was found to be 2.00 mol (cases B5 and C5), while the minimum yield was observed to be 0.95 mol (case C1) It was observed that the syngas yield for the SG cases (A1, B1, C1) decreased with increase in SCR; but the syngas yield was almost constant (approximately 2.00 mol) for DG cases (A5, B5, C5), while the syngas yield slightly decreased for equimolar input moles of H_2O and CO_2 with increase in GaCR from 1 to 3 (cases A3, B3, C3). Similarly, the trends of the individual H_2 and CO gases were also analysed and presented in Figure 3.

Methane

Methane is not a desirable product of gasifier, but is inevitably formed in the gasification process (where steam is in the input). As depicted from Figure 3, the methane yield in CG generally decreased with increase in feed CO_2 moles at constant GaCR (except for A1–A5 conditions where it was almost zero). The methane yield slightly increased for equimolar input moles of H_2O and CO_2 with increase in GaCR from 1 to 3 (cases A3, B3, and C3). As a general observation, the methane yield was insignificant in the gasification product gas at the conditions (1 bar process pressure) chosen in the process design study.

Energy Analysis of Gasifier

Coal Gasifier has two main continuous energy demands for consideration: preheating energy (energy to preheat the gasification reactor raw materials: coal, water, and CO_2 to the gasifier temperature) and gasifier enthalpy (energy required for endothermic gasification reactions). The trends in the gasification reactor enthalpy, gasifier feed preheating requirements, and net energy of the gasifier were studied at 1 bar pressure and respective HCCTs for the different feed conditions (A1 to C5) based on the equilibrium compositions obtained in the earlier section and are shown in Figure 4.

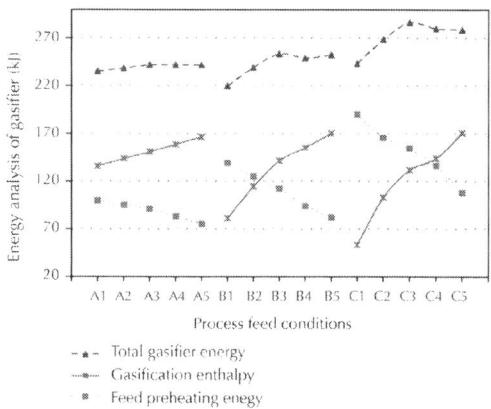

Figure 4: Energies in gasification process.

Gasification Enthalpy

Carbon gasification in absence of air is an endothermic process. It was observed from Figure 4 that the gasification reaction endothermicity generally increased with increase in feed CO_2 moles at constant GaCR for all CG cases. The maximum gasifier enthalpy was found to be 170.86 kJ (case C5), while the minimum reaction enthalpy was observed to be 52.36 kJ (case C1). The gasification enthalpy for the SG cases (A1, B1, and C1) decreased with increase in SCR, while the reaction enthalpy for DG cases (A5, B5, C5) slightly increased with increase in CCR. It was also observed that the gasification enthalpy decreased for equimolar input of H_2O and CO_2 with increase in GaCR from 1 to 3 (cases A3, B3, and C3).

Preheating Energy

As seen in Figure 4, the preheating energy generally decreased with increase in feed CO_2 moles at constant GaCR for all CG cases. The maximum preheating energy was found to be 190.08 kJ for case C1 (highest steam input), while the minimum preheating energy was observed to be 75.22 kJ for case A5. The preheating energy for the SG (cases A1, B1, and C1) increased with increase in SCR and similarly the preheating energy for DG (cases A5, B5, and C5) also increased with increase in CCR. This was due to the huge difference of HCCT of feed conditions. The preheating energy requirement increased for equimolar input of H_2O and CO_2 with increase in GaCR from 1 to 3 (cases A3, B3, and C3), that is, it increased from 90.90 to 154.65 kJ (from A3 to C3).

Total Gasification Energy

The total energy required for gasification is the sum of gasification enthalpy and preheating energy. It was seen that SG requires higher preheating energy, while DG requires higher gasification enthalpy. As seen in Figure 4, the total gasification energy for CG generally increased up to A3, B3, and C3 with increase in feed CO_2 moles (with simultaneous decrease in feed H_2O moles) at constant GaCR and afterwards followed their individual trends. The maximum total energy requirement for gasification was found to be 286.50 kJ (case C3), while the minimum total gasification energy was observed to be 219.99 kJ (case B1). The total energy for the SG (A1, B1, C1) first decreased (due to huge difference of HCCTs) and then increased with increase in SCR, while the total energy for DG cases (A5, B5, C5) only increased with increase in CCR. The total energy for gasification increased for equimolar input of H_2O and CO_2 with increase in GaCR from 1 to 3 (cases A3, B3, and C3).

OC Requirement of CLC Fuel Reactor

The gasifier product gas contains H_2, CO, and CH_4 which are reactive species for CLC fuel reactor. The OC requirement in the CLC fuel reactor depends on the input quantities of these gases which in turn depend on the feed input variations to the gasifier. The product distribution trend of the gasifier product gases has already been discussed in the earlier sections. The stoichiometric requirement of OC (S) for different feed inputs was calculated according to (4), (5), and (6) reactions:

$$H_2 + NiO = H_2O + Ni \qquad (4)$$

$$CO + NiO = CO_2 + Ni \qquad (5)$$

$$CH_4 + 4NiO = CO_2 + 2H_2O + 4Ni \qquad (6)$$

Although these main reactions occur, some side reactions also take place and hence the conversions in the fuel reactor are limited by thermodynamic equilibrium constraints. It was therefore necessary to study the equilibrium product composition of the CLC fuel reactor using stoichiometric amount of OC (Case S) at 1 bar pressure and respective HCCTs. However the OC is generally used in excess of the stoichiometric requirement in CLC processes to enhance the syngas and CH_4 conversion in the fuel reactor. Hence two more cases (1.5S and 2S) using higher amounts (1.5 times and 2.0 times the stoichiometric requirement) of OC were also investigated. Figure 5 shows the syngas and CH_4 compositions in product gases of the CLC fuel reactor for different inputs of OC at various process feed inputs at 1 bar pressure and respective HCCTs. It was observed that the H_2, CO, and CH_4 emissions from the fuel reactor decreased with increase in the amount of OC. It was also observed that the $(CH_4 + CO + H_2)$ moles generally increased with increase in feed CO_2 moles at constant GaCR except for some cases like B4 and C4. The maximum amount of $(CH_4 + CO + H_2)$ exit moles were found to be 0.226 mol (A5), while the minimum quantity of $(CH_4 + CO + H_2)$ exit moles were observed to be 0.127 mol (B1). Higher reactor temperatures (A1–A5) produced relatively higher $(CH_4 + CO + H_2)$ moles for stoichiometric OC usage. The decrease in $(CH_4 + CO + H_2)$ emissions from the CLC fuel reactor was very significant between S and 1.5S conditions, while these emissions were of almost similar magnitude for the use of 1.5S and 2S OC inputs. The complete conversion of syngas and methane to CO_2 and H_2O is highly beneficial in the CLC fuel reactor. Hence the amount of OC to be used in the process was fixed to 1.5S (1.5 times the stoichiometric OC requirement) for further process calculations.

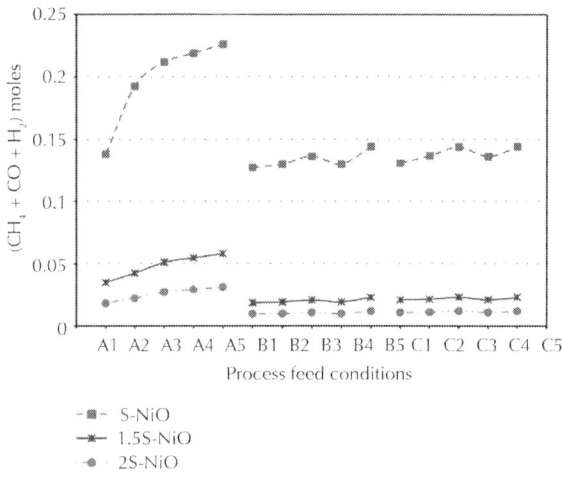

Figure 5: $(CH_4 + CO + H_2)$ moles emitted from the fuel reactor.

Energy Analysis of Fuel Reactor

At a steady state process operation, the fuel reactor receives hot OC from air reactor and hot syngas rich stream from the gasifier. Due to equal reactor temperature assumption, it does not have any feed preheating requirements. The only energy calculations are the fuel reactor enthalpy and energy recoverable from product gases. These fuel reactor energies and the resulting net fuel reactor energies at 1 bar pressure and respective HCCTs were calculated for the different feed conditions (A1 to C5) using the equilibrium compositions obtained in the earlier sections and are plotted in Figure 6.

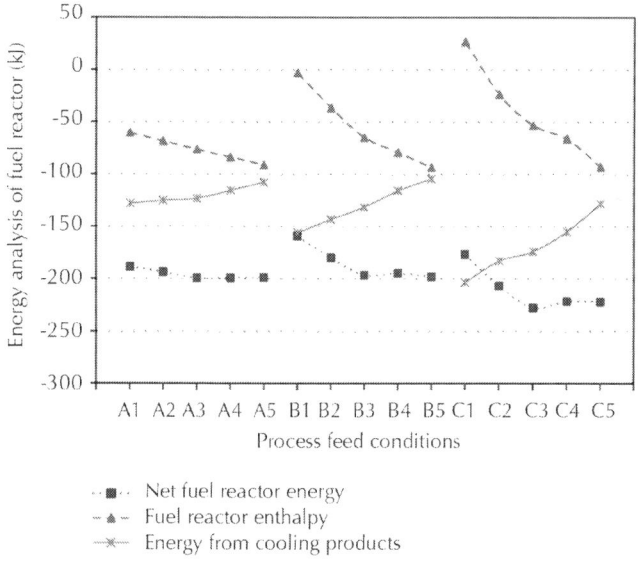

Figure 6: Energies of the CLC fuel reactor.

Fuel Reactor Enthalpy

The fuel reactor enthalpy depends mainly on the syngas and methane content of the gasifier product gas. As seen in Figure 6, the exothermicity of the fuel reactor generally increased with increase in feed CO_2 moles at constant GaCR for all cases. The maximum exothermicity was found to be −93.195 kJ (case B5), while the minimum reactor enthalpy was observed to be 27.44 kJ (case C1). The exothermicity for the SG cases (A1, B1, and C1) decreased with increase in SCR, while in DG cases (A5, B5, C5); the fuel reactor exothermicity was almost constant as the CCR was increased from 1 to 3. It was seen that, the fuel reactor exothermicity decreased for equimolar input of H_2O and CO_2 with increase in GaCR from 1 to 3 (cases A3, B3, and C3).

Extractable Energy by Cooling Fuel Reactor Products

The main gaseous products of the fuel reactor are CO_2 and H_2O at high fuel reactor temperature (except cases A5, B5, and C5 where only CO_2 is emitted by the fuel reactor which can be directly recycled to the gasifier after purging and makeup). For other cases, it was found that minor amount of H_2, CO, and CH_4 are also present in the gaseous streams and the CO_2/H_2O ratio of the fuel reactor product gas was not near to the gasifier CO_2/H_2O ratio requirement. All the 15 streams were considered for heat recovery by cooling to 25°C. It was observed that higher ($H_2 + CH_4$) gasifier stream produced higher steam in fuel reactor which on cooling gave higher extractable energy. As depicted in Figure 6, the extractable energy by cooling products generally decreased with increase in feed CO_2 moles at constant GaCR for all CG cases. The maximum extractable energy by cooling the products was found to be −203.83 kJ (case C1, max. steam input to process), while the minimum extractable energy was seen for case B5 (−104.55 kJ–zero steam input). It was also observed that the extractable energy for the SG cases (A1, B1, C1) increased with increase in SCR (higher steam input gave higher energy on cooling), while in DG cases, that is, A5, B5 and C5, the extractable energy from cooling product gases first slightly decreased from −108.027 kJ to –104.55 kJ (A5 and B5) and then increased to −128.708 kJ (C5) with gradual increase in CCR. It was also observed that the extractable energy increased for cases of equimolar input of H_2O and CO_2 with increase in GaCR from 1 to 3 (cases A3, B3, and C3).

Net Energy of Fuel Reactor

It was observed from Figure 6 that the net energy of the fuel reactor was in the exothermic region for all cases (due to syngas feed). The net energy exothermicity of the fuel reactor in CG cases increased with increase in CO_2 moles till A3, B3, and C3 for respective GaCR and then followed individual trends as shown in the figure. It was also observed that the range of net energy increased with increase in GaCR from 1 to 3. The maximum exothermicity was found to be −228.06 kJ (case C3), while the minimum exothermicity was observed to be −158.34 kJ (case B1).

Material Balance for Air Reactor

The air reactor continuously receives hot depleted OC and carbon from the fuel reactor; unconverted carbon from the gasifier and preheated air to oxidize them to NiO and CO_2 completely. In this study, the unconverted carbon from the gasifier and carbon formed in the fuel reactor were negligible (due to choice of process conditions). The air supply was in exact stoichiometric requirement for the Ni oxidation for its complete conversion to NiO in this study. Figure 7 shows the inputs and outputs of the air reactor at different feed conditions at 1 bar pressure and HCCTs. The input streams to the air reactor are I1-air, I2-Ni, and I3-NiO, while the output streams of the air reactor are O1-NiO and O2-nitrogen. It was observed that there were only slight variations in each stream for the 15 feed conditions considered in this study and hence detailed analysis was not done for this part of study.

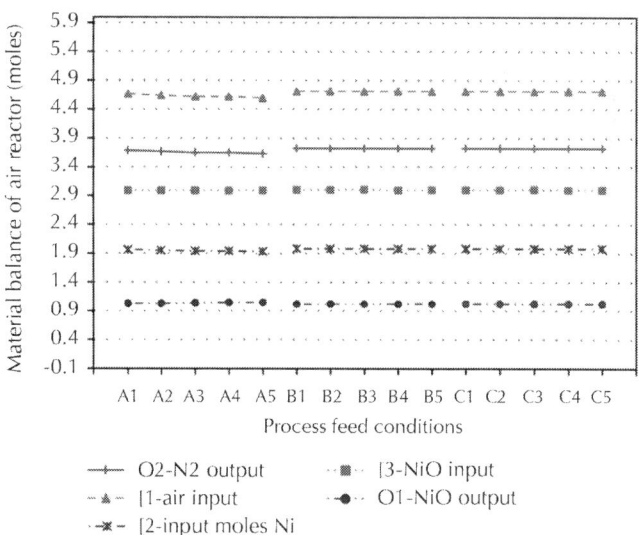

Figure 7: Input and output species of air reactor.

Energy Analysis of Air Reactor

The material balances of the air reactor (discussed in the earlier section) were used to calculate the energy of the air reactor streams. Figure 8 shows the trends in air preheating energy requirement, air reactor enthalpy, energy extracted by cooling of nitrogen, and resulting net energy of air reactor at 1 bar pressure and HCCT for the different feed conditions (A1 to C5). It was seen that the air requirements for the individual 15 cases were almost similar and hence the HCCTs dominated the energy calculations.

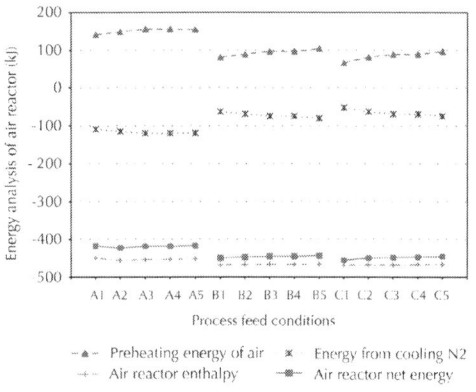

Figure 8: Energies of CLC air reactor.

Preheating Energy Requirement

The air reactor receives preheated air for OC regeneration. As seen in Figure 8, the preheating energy generally increased with increase in feed CO_2 moles at constant GaCR for all CG cases. It was also observed that the air preheating energy requirement generally decreased with increase in combined GaCR due to relatively lower HCCTs. The preheating energy of air for the SG cases (A1, B1, and C1) decreased with increase in SCR and similar observation was noted for DG cases (A5, B5, C5). The preheating energy of air decreased for equimolar input of H_2O and CO_2 with increase in GaCR from 1 to 3 (cases A3, B3, and C3). The maximum preheating energy of air was found to be

155.41 kJ (A3) at 1100°C and minimum preheating energy requirement of air was found to be 67.016 kJ (C1) at 500°C.

Air Reactor Enthalpy

Energy generation in the air reactor is the most important aspect for the process as the air reactor supplies energy for the entire process. It was observed from Figure 8 that the exothermicity of the air reactor slightly varied for all subcases of constant GaCR and it slightly increased with increase in GaCR from 1 to 2 but was almost constant for GaCR 2 and 3 due to nearby individual case HCCTs. The exothermicity of air reactor increased for equimolar input of H_2O and CO_2 with increase in GaCR from 1 to 3 (cases A3, B3, and C3), that is, it increased from −453.56 kJ (A3) to −466.69 kJ (B3) till −468.25 kJ (C3). The maximum exothermicity of the air reactor was found to be −468.88 kJ (C1) at 500°C, while the minimum exothermicity was observed for case A1 (−449.72 kJ) at 1100°C.

Energy Extracted by Cooling Nitrogen

Oxygen from air is used up in the air reactor for regeneration of depleted OC and carbon oxidation. The left-over hot nitrogen gas evolving out of the air reactor can be cooled to 25°C and energy can be extracted for use in the process. It was seen in Figure 8 that this extracted energy generally increased with increase in feed moles at constant GaCR except for case A3, A4, and A5 where it remained almost constant. It was also seen that the exothermicity for SG cases (A1, B1, and C1) and also for DG cases (A5, B5, C5) decreased with increase in respective GaCR due to relatively lower HCCTs. It was also observed that the exothermicity decreased for equimolar input of H_2O and CO_2 with increase in GaCR from 1 to 3 (cases A3, B3, and C3). The maximum extractable energy by cooling nitrogen from the air reactor was found to be −120.39 kJ (A3 and A4 −1100°C) while minimum was found to be −52.075 kJ (C1 −500°C).

Net Energy of Air Reactor

The net energy of air reactor was found by summing the energies of air preheating, air reactor enthalpy and nitrogen stream cooling energy at respective conditions and is shown in Figure 8. It was observed that the trend of net energy of air reactor was similar to that of air reactor enthalpy already discussed in the earlier section. The maximum exothermicity of air reactor was found to be −455.128 kJ (case C1) at 500°C, while the minimum was observed to be −417.22 kJ (case A5). It was seen that the exothermicity of air reactor for DG cases (A5, B5, C5) increased with increase in CCR while the exothermicity of air reactor increased with increase in GaCR from 1 to 3 (cases A3, B3, and C3) for equimolar input of H_2O and CO_2.

Net Process Energy

Net energy of the process is the sum of individual energies of process components (gasifier, fuel, and air reactor). The net energy obtainable from the process at 1 bar pressure and respective HCCTs were calculated for the different feed conditions (A1 to C5) and are shown in Figure 9. The maximum exothermicity for the whole process was found to be −390.157 kJ for case C3 (650°C), while the minimum exothermicity was found to be −372.82 kJ for case A1 at 1000°C. The net energy obtainable from the process was always low for GaCR = 1 cases (A1–A5) due to relative high HCCTs. The preferred conditions for process operation ranged as follows: C3 (−390.157 kJ), C1 (−389.07 kJ), C5 (−388.457 kJ), B5 (−388.43 kJ), B3 (−388.349 kJ), B2 (−388.341 kJ), C2 (−388.24 kJ), B1 (−388.165 kJ), C4 (−388.16 kJ), and B4 (−388.02 kJ). The difference in energies looks small as the calculations are based on 1 mol coal feed but these differences will become huge for a pilot plant operation. A detailed comparison of process energy calculations of gasification coupled CLC and direct air oxidation of coal was also done and is presented in Table 3. The methodology for the direct oxidation calculations was the same that is, preheating carbon and air to reaction temperature (HCCT), combustion enthalpy, energy obtainable by cooling product CO_2 and N_2, and net process energy. It was observed that the energy obtainable in direct coal combustion was slightly higher (3–6 kJ) than the corresponding cases of gasification coupled CLC. Thus the magnitude of net energy obtainable in gasification coupled CLC process and direct coal combustion process was found to be of similar nature. But the gasification coupled CLC process delivered an almost pure CO_2 stream for sequestration making it clean energy process.

Table 3: Comparison of net energy obtainable from direct combustion and gasification coupled CLC of coal

Case	C3	C1	C5	B5	B3	B2	C2	B1	C4	B4
HCCT (°C)	650	500	700	750	700	650	600	600	650	700
Air required (moles)	3.76	3.76	3.76	3.76	3.76	3.76	3.76	3.76	3.76	3.76
Preheating of coal (kJ)	10.06	7.09	11.11	12.19	11.11	10.06	9.04	9.04	10.06	11.11
Energy to heat air (kJ)	90.20	67.69	97.80	105.45	97.80	90.20	82.65	82.65	90.25	97.80
Reaction enthalpy from direct combustion (kJ)	−394.44	−394.11	−394.54	−394.65	−394.54	−394.44	−394.33	−394.33	−394.44	−394.54
Energy to cool N_2 (kJ)	−69.95	−52.60	−75.81	−81.71	−75.81	−69.95	−64.13	−64.13	−69.95	−75.81
Energy to cool CO_2 (kJ)	−29.56	−21.78	−32.22	−34.91	−32.22	−29.56	−26.93	−26.93	−29.56	−32.22
Net energy from direct combustion (kJ)	−393.68	−393.71	−393.66	−393.63	−393.66	−393.68	−393.69	−393.69	−393.68	−393.66
Net energy from gasification coupled CLC process (kJ)	−390.16	−389.07	−388.46	−388.43	−388.35	−388.34	−388.24	−388.17	−388.16	−388.02
Energy Difference(kJ)	3.52	4.64	5.20	5.20	5.31	5.34	5.46	5.53	5.52	5.64

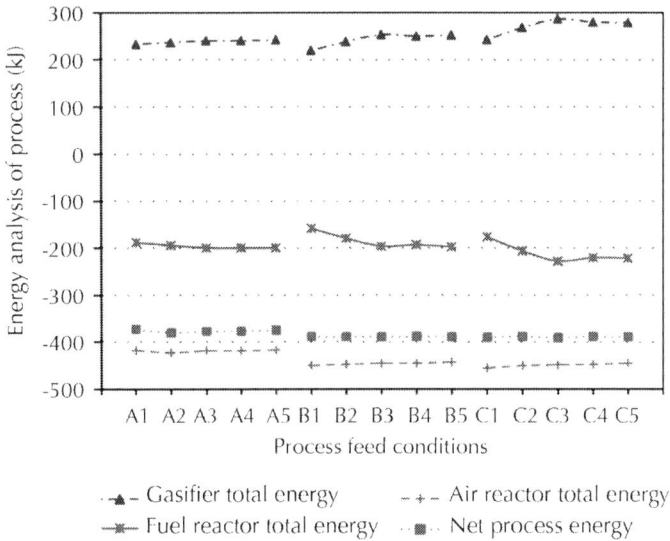

Figure 9: Trend of net process energy.

CONCLUSIONS

This theoretical systematic process study was done to understand the material and energy balances of the new process design of gasification coupled CLC of coal. The comprehensive study considered the effect of temperature, pressure and combined gasification using steam and CO_2 on the process in steps. It was concluded that coal (carbon) gasification using 1.5 mol H_2O and 1.5 mol CO_2 per mole carbon at 1 bar pressure and 650°C delivered maximum energy from the process. The other process conditions yielding near maximum energies are also identified and can be used as necessary. The results obtained in this detailed study can be used for scale up of the process. Experimental evaluation will be helpful to further enhance the technology commercialization prospect. Use of gasification step in this study can also help design similar processes using other fuels such as natural gas, which can be converted to syngas (via steam/dry reforming) and then used in CLC processes. The CLC processes may ultimately become limited to syngas-CLC, which will help the OC development and reactor design aspects and ensure fast track commercialization. This process scheme has an exothermic syngas fuel reactor which may help solve many heat transfer problems in the CLC systems. Further studies using this process design and actual conditions like coal compositions, incomplete coal conversion in the gasifier, process heat losses, different reactor operating temperatures and pressures, operating energies of intermediate process equipments, for example, gas solid separators, and so forth can also be evaluated to generate data to help design practical coal CLC systems.

REFERENCES

1. E. Shoko, B. McLellan, A. L. Dicks, and J. C. D. da Costa, "Hydrogen from coal: production and utilisation technologies," International Journal of Coal Geology, vol. 65, no. 3-4, pp. 213–222, 2006. · ·
2. G. Tsatsaronis, K. Kapanke, and A. M. B. Marigorta, "Exergoeconomic estimates for a novel zero-emission process generating hydrogen and electric power," Energy, vol. 33, no. 2, pp. 321–330, 2008. · ·
3. A. Hesenov, H. Kinik, G. Puli, B. Gözmen, S. Irmak, and O. Erbatur, "Electrolysis of coal slurries to produce hydrogen gas: relationship between CO_2 and H2 formation," International Journal of Hydrogen Energy, vol. 36, no. 9, pp. 5361–5368, 2011. · ·
4. S. Lin, M. Harada, Y. Suzuki, and H. Hatano, "Hydrogen production from coal by separating carbon dioxide during gasification," Fuel, vol. 81, no. 16, pp. 2079–2085, 2002. · ·
5. M. Abdollahi, J. Yu, K. T. L. Paul, R. Ciora, M. Sahimi, and T. T. Tsotsis, "Hydrogen production from coal-derived syngas using a catalytic membrane reactor based process," Journal of Membrane Science, vol. 363, no. 1-2, pp. 160–169, 2010.
6. A. Martinez, K. Gerdes, G. Randall, and J. Poston, "Thermodynamic analysis of interactions between Ni-based solid oxide fuel cells (SOFC) anodes and trace species in a survey of coal syngas," Journal of Power Sources, vol. 195, no. 16, pp. 5206–5212, 2010. · ·

7. S. Ghosh and S. De, "Energy analysis of a cogeneration plant using coal gasification and solid oxide fuel cell," Energy, vol. 31, no. 2-3, pp. 345–363, 2006.
8. R. S. Christopher and M. Alejandro, "Fundamental investigation of NOx formation during oxy-fuel combustion of pulverized coal," Proceedings of the Combustion Institute, vol. 33, pp. 1723–1730, 2011.
9. M. Hishida, M. Fumizawa, Y. Inaba et al., "Nuclear energy conversion systems for arresting global warming," Energy Conversion and Management, vol. 38, no. 10–13, pp. 1365–1375, 1997. · ·
10. Wmo Greenhouse Gas Bulletin, No 7, World Meterological Organization, 2011.
11. A. Abad, T. Mattisson, A. Lyngfelt, and M. Rydén, "Chemical-looping combustion in a 300 W continuously operating reactor system using a manganese-based oxygen carrier," Fuel, vol. 85, no. 9, pp. 1174–1185, 2006. · ·
12. H. Jin and M. Ishida, "A new type of coal gas fueled chemical-looping combustion," Fuel, vol. 83, no. 17-18, pp. 2411–2417, 2004. · ·
13. A. Lyngfelt, B. Leckner, and T. Mattisson, "A fluidized-bed combustion process with inherent CO_2 separation; application of chemical-looping combustion," Chemical Engineering Science, vol. 56, no. 10, pp. 3101–3113, 2001. · ·
14. J. S. Dennis, C. R. Müller, and S. A. Scott, "In situ gasification and CO_2 separation using chemical looping with a Cu-based oxygen carrier: performance with bituminous coals," Fuel, vol. 89, no. 9, pp. 2353–2364, 2010. · ·
15. M. M. Hossain and H. I. de Lasa, "Chemical-looping combustion (CLC) for inherent CO_2 separations-a review," Chemical Engineering Science, vol. 63, no. 18, pp. 4433–4451, 2008. · ·
16. Y. Cao and W. P. Pan, "Investigation of chemical looping combustion by solid fuels. 1. Process analysis,"Energy and Fuels, vol. 20, no. 5, pp. 1836–1844, 2006.
17. T. Mattisson, A. Lyngfelt, and H. Leion, "Chemical-looping with oxygen uncoupling for combustion of solid fuels," International Journal of Greenhouse Gas Control, vol. 3, no. 1, pp. 11–19, 2009. · ·

18. W. Xiang, S. Wang, and T. Di, "Investigation of gasification chemical looping combustion combined cycle performance," Energy and Fuels, vol. 22, no. 2, pp. 961–966, 2008. · ·
19. M. Zheng, L. Shen, and J. Xiao, "Reduction of $CaSO_4$ oxygen carrier with coal in chemical-looping combustion: effects of temperature and gasification intermediate," International Journal of Greenhouse Gas Control, vol. 4, no. 5, pp. 716–728, 2010. · ·
20. L. Shen, J. Wu, and J. Xiao, "Experiments on chemical looping combustion of coal with a NiO based oxygen carrier," Combustion and Flame, vol. 156, no. 3, pp. 721–728, 2009. · ·
21. X. Wang, B. Jin, W. Zhong, Y. Zhang, and M. Song, "Three-dimensional simulation of a coal gas fueled chemical looping combustion process," International Journal of Greenhouse Gas Control, vol. 3, pp. 1750–5836, 2011.
22. J. Corella, J. M. Toledo, and G. Molina, "Steam gasification of coal at low-medium (600-800°C) temperature with simultaneous CO_2 capture in a bubbling fluidized bed at atmospheric pressure. 2. Results and recommendations for scaling up," Industrial and Engineering Chemistry Research, vol. 47, no. 6, pp. 1798–1811, 2008. · ·
23. J. Matsunami, S. Yoshida, Y. Oku, O. Yokota, Y. Tamaura, and M. Kitamura, "Coal gasification by CO_2 gas bubbling in molten salt for solar/fossil energy hybridization," Solar Energy, vol. 68, no. 3, pp. 257–261, 2000. ·
24. M. F. Irfan, M. R. Usman, and K. Kusakabe, "Coal gasification in CO_2 atmosphere and its kinetics since 1948: a brief review," Energy, vol. 36, no. 1, pp. 12–40, 2011. · ·
25. D. P. Ye, J. B. Agnew, and D. K. Zhang, "Gasification of a South Australian low-rank coal with carbon dioxide and steam: kinetics and reactivity studies," Fuel, vol. 77, no. 11, pp. 1209–1219, 1998. ·
26. A. Molina and F. Mondragón, "Reactivity of coal gasification with steam and CO_2," Fuel, vol. 77, no. 15, pp. 1831–1839, 1998. ·
27. C. Nenad, R. Branislav, M. Rastko, N. Olivera, and V. Miomir, "Experimental investigation of role of steam in entrained flow coal gasification," Fuel, vol. 86, no. 1-2, pp. 194–202, 2007. · ·

28. A. Abad, F. G. Labiano, L. F. de Diego, P. Gayán, and J. Adánez, "Reduction kinetics of Cu-, Ni-, and Fe-based oxygen carriers using syngas (CO + H_2) for chemical-looping combustion," Energy and Fuels, vol. 21, no. 4, pp. 1843–1853, 2007. · ·
29. T. Mattisson, F. G. Labiano, B. Kronberger, A. Lyngfelt, J. Adánez, and H. Hofbauer, "Chemical-looping combustion using syngas as fuel," International Journal of Greenhouse Gas Control, vol. 1, no. 2, pp. 158–169, 2007. · ·
30. S. P. Singh, S. A. Weil, and S. P. Babu, "Thermodynamic analysis of coal gasification processes," Energy, vol. 5, no. 8-9, pp. 905–914, 1979. ·
31. M. Gassner and F. Maréchal, "Thermodynamic comparison of the FICFB and Viking gasification concepts," Energy, vol. 34, no. 10, pp. 1744–1753, 2009. · ·
32. Z. Wang, J. Zhou, Q. Wang, J. Fan, and K. Cen, "Thermodynamic equilibrium analysis of hydrogen production by coal based on Coal/CaO/H_2O gasification system," International Journal of Hydrogen Energy, vol. 31, no. 7, pp. 945–952, 2006. · ·
33. M. Díaz-Somoano and M. R. Martínez-Tarazona, "Trace element evaporation during coal gasification based on a thermodynamic equilibrium calculation approach," Fuel, vol. 82, no. 2, pp. 137–145, 2003. · ·
34. A. Dufaux, B. Gaveau, R. Létolle, M. Mostade, M. Noël, and J. P. Pirard, "Modelling of the underground coal gasification process at Thulin on the basis of thermodynamic equilibria and isotopic measurements," Fuel, vol. 69, no. 5, pp. 624–632, 1990. ·
35. X. Li, J. R. Grace, A. P. Watkinson, C. J. Lim, and A. Ergüdenler, "Equilibrium modeling of gasification: a free energy minimization approach and its application to a circulating fluidized bed coal gasifier," Fuel, vol. 80, no. 2, pp. 195–207, 2001. · ·
36. A. Melgar, J. F. Pérez, H. Laget, and A. Horillo, "Thermochemical equilibrium modelling of a gasifying process," Energy Conversion and Management, vol. 48, no. 1, pp. 59–67, 2007. · ·
37. Y. Cao and W. P. Pan, "Investigation of chemical looping combustion by solid fuels. 1. Process analysis," Energy and Fuels, vol. 20, no. 5, pp. 1836–1844, 2006.

38. J. P. E. Cleeton, C. D. Bohn, C. R. Müller, J. S. Dennis, and S. A. Scott, "Clean hydrogen production and electricity from coal via chemical looping: identifying a suitable operating regime," International Journal of Hydrogen Energy, vol. 34, no. 1, pp. 1–12, 2009. · ·
39. Z. P. Gao, L. Shen, J. Xiao, C. Qing, and Q. Song, "Use of coal as fuel for chemical-looping combustion with Ni-based oxygen carrier," Industrial & Engineering Chemistry Research, vol. 47, pp. 9279–9287, 2008.
40. HSC Chemistry [software], Version 5. 1. Pori, Outokumpu Research Oy, 2002.
41. W. R. Smith, "Computer software reviews," Journal of Chemical Information and Computer Sciences, vol. 36, no. 1, pp. 151–152, 1996, HSC Chemistry for Windows, 2.
42. R. H. Perry and D. W. Green, Chemical Engineers›Handbook, McGraw-Hill, 7th edition, 1997.

Chapter 4

Analysis of Changes in the Properties of Selected Chemical Compounds and Motor Fuels Taking Place during Oxidation Processes

Marta Skolniak[1], Paweł Bukrejewski[1], and Jarosław Frydrych[1]

[1]Oil Products and Biofuels Laboratory, Automotive Industry Institute, Poland

INTRODUCTION

Given the existing energy crisis and restrictions on GHG emissions, it has become a necessity to introduce to motor fuels newer and newer types of non-petroleum components – especially biocomponents such as ethanol or FAME.

The path of the oxidation (ageing) process for fuels which contain biocomponents as additives has not been determined so far.

Under the circumstances, the ever more increasing demand on liquid fuels obtained by petroleum processing, combined with the ever more stringent quality requirements, it has become an important aspect to maintain a high quality of fuels during long-term storage. The different physical and chemical characteristics observed in petroleum products after storing them for a long time depend mainly on the chemical composition of petroleum, 90% of which is a mixture of hydrocarbons having different structures.

Petroleum contains the following types of hydrocarbons [1]:
- paraffins (n-paraffins, isoparaffins),
- olefins,
- cycloparaffins (naphthenes),
- aromatic hydrocarbons.

Paraffin hydrocarbons (paraffins) are present in large amounts in petroleum and they predominate in large amounts in gasoline fractions. They include straight chained n-paraffins and branched isoparaffins. n-Paraffins are low-reactivity compounds, therefore, they are applicable in the refinery and other industries. Isoparaffins are also found in petroleum but the number of potential compounds is immense. Isoparaffins have lower boiling points, compared with n-paraffins.

Naphthene hydrocarbons (naphthenes) are the largest fraction in petroleum. A naphthene ring contains typically 5 or 6 carbon atoms. Dicyclonaphthenes C8 and C9 and (more condensed) cycloparaffins, are present in addition to monocyclonaphthenes. Cycloalkanes are less volatile compounds, compared with alkanes.

Aromatic hydrocarbons (aromatics) have a ring-like structure. The benzene ring comprises alternate double and single bonds with adjacent carbons atoms. Monocyclic aromatics are valuable components in fuels, especially gasolines, because they improve their octane number. Aromatic hydrocarbons are highly stable at high temperatures, thereby leading to the desirable effect of knock-less combustion in gasoline-fueled engines. They are not desirable in fuels intended for use in spontaneous-ignition engines because they cause rough engine operation and decrease the fuel's cetane number, thus delaying spontaneous ignition.

In addition to the aforementioned groups of hydrocarbons, petroleum also comprises non-hydrocarbon ingredients of which the molecules contain atoms of sulfur, nitrogen, and oxygen, organometallic bonds and inorganic salts.

The following types of motor fuels are available in the market:
- motor fuels for use in vehicles equipped with spark-ignition engines,
- diesel fuel for use in vehicles equipped with spontaneous-ignition engines.

Motor gasolines are a mixture of organic compounds which boil in the temperature range from 30 to 200°C [1, 3]. The mixture contains components from the paraffins, naphthenes, olefins, and aromatic hydrocarbon groups with C4 to C10 carbon atoms per molecule. As used currently, motor gasolines also have a content of non-hydrocarbon components, such as ethers, and alcohols, as well as additives which improve the motion performance of fuels. The content of the various hydrocarbon fractions in gasoline may vary widely for different types of raw material, refining technology, process conditions, and qualitative requirements.

Most frequently, the streams used for blending motor gasolines are derived from such processes as:
- distillation of petroleum,
- alkylation, isomerization, and solvent extraction,
- thermal cracking,
- catalytic cracking,
- catalytic reforming,
- hydrocarbon processes (hydrorefining, hydrocracking),
- other processes.

A gasoline from vacuum-oil catalytic cracking comprises a large fraction of isoparaffins, olefins, and aromatics (aromatic hydrocarbons). A product of reforming (reformate) has a high content of aromatics. A product of isomerization (isomerizate) contains essentially 2, 2-dimethylbutane and isopentane as well as certain amounts of 2-methylpentane and 2, 3-dimethylbutane. Polymerization gasoline obtained from C3 and C4 olefins is an olefin product, while an alkylate is a component with a zero content of aromatics, olefins or benzene.

In addition to the hydrocarbon fractions, gasolines may have a content of oxygen components such as alcohols (including methyl, ethyl, isopropyl, and isobutyl alcohols), butyl ether as well as methyl *tert*-butyl ether, ethyl *tert*-butyl ether, methyl *tert*-amyl ether. Other alcohols may also be present although their presence is limited by their boiling point (max. 215°C).

In the storage, transport, and handling processes during the distribution train, motor gasolines are affected by physical and chemical factors, whereby their physical and chemical processes are changed. As the result, the motor gasoline in one's vehicle's fuel tank may have different properties, compared with those of a fresh made product from the refinery: its quality may be deteriorated. Its properties may be affected by the following phenomena:

- evaporation of light ends during the handling (filling) and storage of gasoline,
- the effect of atmospheric oxygen, leading to the occurrence of oxidation and polymerization of certain components of the fuel,
- penetration of impurities (particles) into gasoline,
- penetration of atmospheric moisture into gasoline.

Oxidation and polymerization of motor gasoline take place usually during long-term storage in storage tanks and in vehicle fuel tanks. The intensity of such processes is higher in fuels with a higher percentage of reactive components, increased contact with oxygen (less fuel in the tank), and at higher temperatures of the fuel itself. Such processes lead to the formation of asphalts and gums which are suspended in the engine gasoline along with corrosion products and mineral particles, and may also precipitate in the form of deposits.

Diesel fuels are a mixture of hydrocarbons with C11-C25 carbon atoms per molecule and a boiling range from 150 to 400°C [1, 2]. The properties of the fuels intended for use in spontaneous-ignition engines are much different from those of motor gasoline. Diesel fuels consist of the following hydrocarbon groups:

- n-paraffins : 9-13% (V/V)
- isoparaffins: 30-55% (V/V)
- naphthenes: 25-35% (V/V)
- aromatics: : 15-30% (V/V)
- olefins: : 0-5% (V/V)

Paraffin hydrocarbons have between 10 and 20 carbon atoms per molecule. They improve the fuel's cetane number while deteriorating its low-temperature properties.

Cycloparaffins (also called naphthenes) in diesel fuels are mainly alkylcyclohexanes, decahydronaphthalenes and perhydronaphthalenes. The percentage of that group of compounds depends on what type of petroleum is processed and on how much of the diesel fuel fraction originates from catalytic cracking.

Aromatic hydrocarbons include alkylbenzenes, indanes, naphthalenes, biphenyls, acenaphthenes, phenanthrenes, chrysenes and pyrenes. Among those compounds, the highest percentage is that of naphthalenes.

Crude components of diesel fuel originate from the following processes:
- atmospheric distillation of petroleum,
- catalytic cracking of various petroleum fractions,
- thermal cracking of various petroleum fractions,
- hydrocracking of distillates or distillation bottoms,
- vacuum distillation,
- and other ones.

The principal contaminants of diesel fuel include:
- asphaltenes and gums (they are formed by oxidation and polymerization of reactive components of diesel fuel (unsaturated hydrocarbons, sulfur, nitrogen, and oxygen compounds),
- sulfur and sulfur compounds which were not removed in the production process,
- dust particles,
- corrosion products from tanks and pipelines,
- water,
- microorganisms and their metabolites.

As in the case of gasoline, oxidation and polymerization of diesel fuel take place during its long-term storage, when the fuel is in contact with oxygen or contains reactive components, or when the storage temperature is too high. Such processes produce asphalts and gums which form sludge-type deposits. The formation of such deposits is induced by diesel fuel oxidation products, corrosion and dust particles.

This chapter is intended to provide an answer to some of the important issues connected with the effect of the chemical structure of organic components of fuels on the oxidation stability of motor gasolines and diesel fuels which are used in motor transport.

RESEARCH METHODOLOGY

The Selected Motor-fuel Ageing Method

The principal hydrocarbon components of fuels and their derivatives which occur or potentially occur in fuels were selected for the examination of fuel ageing processes, and their stability was determined. The compounds were oxidized in accordance with EN 16091 in order to find the time of stability of the respective fuel components. The resulting products were then analyzed using infrared spectrometry and gas chromatography with mass detector.

The selected accelerated fuel ageing test enables evaluation of the oxidation stability of test products in a short time. Only a small fuel volume is required for the test (approximately 10 ml was used in this case). The test is safe and is controlled using a microprocessor. The automatic control process covers heating, cooling, as well as rinsing, and filling with oxygen. Pressure drop is measured and recorded using a suitable microprocessor with a highly sensitive pressure sensor; the data are transmitted to a computer using an interface for further processing and the final result is a diagram which shows the pressure vs. time relationship. The sample preparation process is highly automated and the sensors function very precisely. Therefore, information about the oxidation process pathway shows that the test method is highly repeatable, compared with currently used test methods.

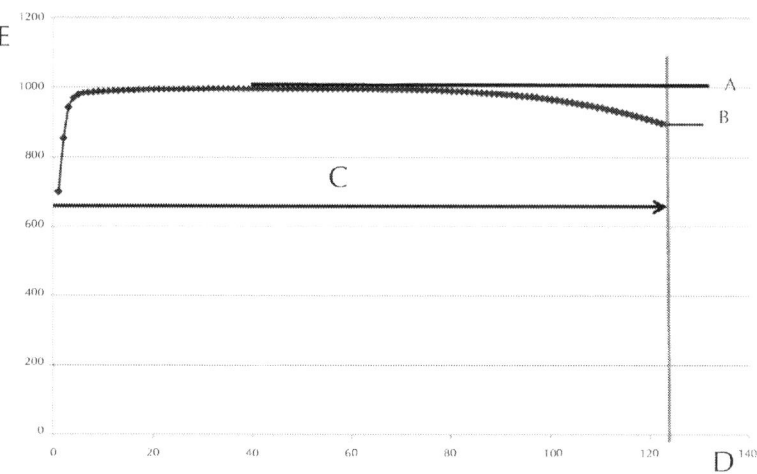

Figure 1: The rapid-oxidation test course diagram, A – the maximum pressure recorded, P_{max}, B – breakpoint, C – induction period (min), D – duration of test (min), E – pressure value recorded, kPa.

The test method is based on the measurement of pressure changes in a tightly-sealed test vessel. The test vessel (made of metal and coated with a thin coat of gold) is filled with 10 ml of the test sample. After being tightly covered with a lid, the test vessel is rinsed with oxygen to remove any air present in the space above the test sample. The test vessel is then filled with oxygen to obtain a pressure of 500 kPa (at ambient temperatures). The pressurized vessel, filled with the test sample and oxygen, is heated to 140°C; the temperature increase causes pressure buildup. The test conditions (temperature and pressure) are stabilized within about 2 minutes. The period of time of pressure stabilization is strictly related to the sample's oxidation stability. Fuel samples with low oxidation stabilities will soon become oxidized throughout their volumes, leading to a considerable pressure drop in the closed vessel. The time of pressure stabilization for samples with high oxidation stabilities will be much longer. Temperature and pressure in the vessel are recorded at 1 sec intervals until the end point is reached, that is, the pressure drops by 10% of its highest value. The test measure is the time which has lapsed between the test commencement (that is, the time when the sample reached 140°C) and the time when pressure inside the test vessel dropped by 10%. Figure 1 shows the characteristics of the course of the oxidation stability test for a motor fuel.

Selected Analytical Methods

Chromatography with Mass Spectrometry (GC MS)

Among all chromatographic techniques known in the art, gas chromatography provides unsurpassed resolutions. Therefore, the technique is most frequently combined with the spectroscopic technique for a fast identification of the separated compounds. Mass spectroscopy is most typically used for the purpose.

In the chromatographic technique, the mixture separation takes place on the chromatographic column. Three types of chromatographic columns are known: column with solid adsorbent, columns with solid-liquid adsorbent, and capillary columns. The capillary columns are typically used in gas chromatography.

Detector is a major component in every gas chromatograph. It is the part all the substances flow into after being separated on the chromatographic column. The ideal detector is sensitive only to the concentration of a compound, regardless of its chemical structure. However, detectors have different sensitivities to various chemical compounds, therefore, its is necessary to calibrate detectors and establish what is called "response factor" separately for every chemical compound to be able to determine with good precision the percentages of the various chemical compounds contained in the analyzed sample.

The Electron Ionization Detector (EID), also called the Mass Selective Detector (MSD), provides information about the analyzed compound in the form of its electron ionization mass spectrum. The spectrum is characteristic of every chemical compound, except certain isomers – they may have same spectra [4].

The interface is an important part, connecting the gas chromatograph with the mass spectrometer. While the pressure at the outlet of the chromatographic column is atmospheric, the next component of the system (ionization chamber of the mass spectrometer) typically operates at pressures in the range 10^{-4}-10^{-3} Pa. It is the primary task of the interface to form a connection which provides optimum operating conditions for the two parts.

Ions from the ionization chamber are introduced into the mass analyzer, where they are separated according to their mass-to-load ratios and, since the load is usually equal to 1, then separation takes place by the mass value. Out of a number of types of analyzers, the following ones are most suitable for the connected GC-MS system: quadrupole analyzer (so called "ion trap"), magnetic analyzer, and time-of-flight analyzer.

The operation of a detector is based on substance introduction – from the chromatographic column – into the vacuum chamber, where the substance is ionized. The ions being formed are focused and accelerated in the mass filter. The mass filter selectivity enables the passage of all ions having a certain mass into the electron multiplier. All of the ions having that certain mass are detected. The mass filter then enables the passage of another mass which is different from the mass of other ions. The preset mass range is scanned by the mass filter gradually, several times a second. The total number of ions is counted every time. The ion intensity or number after each scan is plotted vs. time in the chromatogram (for TIC – total ionic current). A mass spectrum is obtained for each scan; it indicates various ion masses vs. their intensity or number.

The analyses were performed using a GC-MS apparatus from Agilent. The apparatus is equipped with a non-polar HP-5MS column having the following parameters: length: 30 m, diameter: 0.25 mm, film thickness: 0.25 mm, and packing: (5%-phenyl)-methylpolysiloxane. The operating parameters of the apparatus were as follows:
- injector temperature: 250°C (optionally for heavier components 300°C, injection volume: 0.2 ml, stream split: 1:50,
- oven temperature program: 40°C (70°C optionally for heavier components) – 4 min, 10°C/min to a temperature of 180°C (305°C),
- flow of carrier gas (He) – 1 ml/min,
- ion source temperature: 230°C.

Infrared Spectrometry

Matter is able to interact with radiation through absorption or emission. The two processes are based on photon absorption or emission through a particle of matter; the photon energy corresponds to the energy difference between the initial and final states of the molecule: in the case of absorption, the final state is the one having a higher energy, compared with the initial state; in the case of emission, the energy of the final state is lower than that of the initial state: the difference indicates the energy of the absorbed or emitted photon, respectively. Infrared spectroscopy measures the absorption of infrared radiation by the molecules of chemical compounds [5].

Infrared is the range of radiation with a wavelength from 780 nm (a conventional end point of the visible range) to 1 mm (a conventional start point of the microwave range). In practice, the medium infrared range from 2.5 μm to 25 μm (or from 4000 cm^{-1} to 400 cm^{-1}) is typically applied.

The absorption of infrared radiation for a majority of known particles causes their excitations (passages) onto higher oscillation levels. However, not all passages are active or have measurable intensities. The active passages, also referred to in spectroscopy as permissible passages, must satisfy certain criteria, referred to as the rules of choice. In the infrared, the only active passages are those of polar molecules having non-zero dipole moments. Speaking in more precise terms, only those vibrations may be excited in the molecule which change the dipole moment of that molecule. Moreover, the most active passages exist between the adjacent levels of oscillation for a given vibration.

The total number of vibrations is 3N-6 for a non-linear molecule and 3N-5 for a linear molecule, where N is the number of atoms per molecule. If the molecule has an element or elements of symmetry, then not all vibrations will be shown in the spectrum. Infrared spectroscopy provides information about the test material in the form of a spectrum – a diagram showing absorption vs. energy of radiation, which is usually expressed as the wave number ($\tilde{}$ [cm^{-1}]).

Every molecule has its unique set of energy levels, therefore, infrared spectra are typical of specific chemical compounds. Comparing the spectrum of a given substance with a previously created spectral library is one of the available methods to identify compounds by means of infrared spectroscopy.

Another method is based on the assigning of bands to the vibrations of the specific functional groups present in the molecule of a given chemical compound, using vibration correlation tables. A given functional group (several atoms, connected by means of chemical bonds eg., carbonyl group –C=O, hydroxyl group –OH) occurring in different compounds has similar values of vibration frequency (energy). The observed frequency ranges which are typical of a given group along with its vibrations, have been collected in correlation tables. Table 1 shows the wave numbers for the characteristic absorption of several frequently occurring functional groups.

Table 1: Characteristic wave numbers of bands originating from typically existing bonds in organic compounds

Bond	Type of vibration	Location [cm-1]
O-H (water)	stretching	3760
O-H (alcohols and phenols)	stretching	3650-3200
O-H (carboxylic acids)	stretching	3650-2500
N-H	stretching	3500-3300
C-H (alkynes)	stretching	3350-3250
C-H (vinyl and aryl)	stretching	3100-3010
C-H (aliphatic)	stretching	2970-2850
C N	stretching	2280-2210
C C	stretching	2260-2100
C=O	stretching	1760-1690
C=N	stretching	1750-1500
C=C (alkenes)	stretching	1680-1610
N-H	deformation	1650-1550
C=C (aryl)	stretching	1600-1500
C-C (aliphatic)	stretching	1500-600
C-H (aliphatic)	deformation	1370-1340
C-N	stretching	1360-1180
C-O	stretching	1300-1050
C-H (vinyl)	deformation	995-675
C-H (aryl)	deformation	900-690

Measurement techniques may, essentially, be divided into transmission and reflection techniques. In the transmission techniques, the oscillation spectrum is measured by measuring the radiation intensity after passing through the sample. A drop in intensity for the incident beam indicates absorption of radiation by the sample. Owing to the low transparency of materials in the medium infrared range, the use of the method requires an amount of effort and resourcefulness in preparing the samples. Measurements of the transmission spectra for gases and liquids are carried out using cells with window cells made of materials (such as KBr, NaCl) which are transparent to the infrared range. The spectra of solids can be measured in pellets made of alkali metal halides (KBr), in the form of suspension in Nujol (liquid paraffin), on silicon plates. If the test object is thin enough for radiation to be able to pass through, transmission spectra may be measured directly. In transmission techniques, the measure of absorption of a radiation with a specific wave number ($\tilde{v}$) through the sample may either be transmittance ($T(\tilde{v})$) or absorbance $A(\tilde{v})$, both of which are defined by means of Equations (1) and (2). Absorbance is a practical value: it is useful for the quantitative description of absorption and its value is directly proportional to the number of the absorbing molecules according to the Bouguer-Lambert-Beer rule.

$$T(\tilde{v}) = \frac{I}{I_0} \qquad (1)$$

$$A(\tilde{v}) = \log \frac{I_0}{I} = -\log T \qquad (2)$$

wherein: I_0 – incident beam intensity when falling onto the sample, I – beam intensity after passing through the sample.

Reflection techniques enable the infrared spectra to be obtained by measuring radiation after it is reflected from the sample. The reflected radiation is measured by means of various optical systems used in attachments for spectrometers. The most typical systems are based on total reflection (mirror reflection), attenuated total reflection (ATR) or diffuse reflectance infrared Fourier transformed spectroscopy, DRIFT).

The ATR method is based on the total internal reflection of light. In that phenomenon, the light beam is introduced into a material which is transparent to the infrared and has a high refractive index (eg. for diamond) and falls onto its inner surface. The test sample is pressed to the outer side of that surface at the reflection point. Such radiation is subject to the total internal reflection and will not get outside the medium in which it was moving, although its energy may be absorbed by the sample located on the other side. The beam light is then taken out of the medium where the total inner reflection occurred, thus making it possible to measure its intensity and the infrared spectrum.

The diffuse reflection is the type of reflection where the angle of reflection is different from the angle of incidence. It occurs when the surface roughness is rather significant, compared with the wavelength. The incident radiation may penetrate deep into the sample where it is reflected off the consecutive layers of atom a number of times and is somewhat attenuated, only to leave the sample at a different angle, compared with the angle if incidence. The intensity of radiation which is reflected in a diffuse manner is measured with a system of mirrors or a spherical mirror whereby the radiation, after being reflected in all directions, is directed into the detector. In addition, beam stops are used in order to eliminate part of the radiation which is reflected in the mirror-like manner. The method of diffuse reflectance infrared Fourier transformed spectroscopy (DRIFT) is used for examination of samples in the form of powder or matt surfaces.

In reflection techniques, the absorption measures used are reflectance ($R(\tilde{v})$) and the negative logarithm of reflectance ($-\log R(\tilde{v})$). The values are analogs of transmittance and absorption, used in the case of transmission techniques and are described by the formulas (3) and (4).

$$R(\tilde{v}) = \frac{I}{I_0} \qquad (3)$$

$$-\log R(\tilde{v}) = \log \frac{I_0}{I} \qquad (4)$$

wherein: I_0 – incident beam intensity when falling onto the sample,
I – beam intensity after being reflected from the sample

The spectrometric analysis was carried out using the Magna System 750 spectrometer, equipped with: white light source, a DTGS KBr detector, and KBr cell.

Its operating parameters were as follows:
- number of scans: 32,
- length of measurement: 38.73 s,
- resolution: 4 000.

OXIDATION STABILITY OF PURE COMPOUNDS

Compounds which are typically present both in gasoline and diesel fuel were selected for oxidation stability tests. The following hydrocarbons and their derivatives were selected: n-hexane, 1-hexene, 1-hexyne, n-heptane, n-octane, isooctane, cetane (n-hexadecane), benzene, toluene, cyclohexane, o-xylene, ethanol, methyl t-butyl ether (MTBE), pentanoic (valeric) acid methyl ester.

In addition, a mixture of fatty acid methyl esters (FAME) was subjected to ageing. Findings of the oxidation stability tests for the selected compounds are shown in Figure 2.

The results shown in Figure 2 indicate that oxidation stability is the lowest for the hydrocarbons having multiple bonds and for FAME with a large number of double bonds. High oxidation stability is shown by short-chained hydrocarbons, branched hydrocarbons, aromatics, and ethers.

Aliphatic chain length vs. Oxidation stability

Saturated hydrocarbons with different chain lengths were selected for the oxidation stability tests. Their oxidation stability was tested (Figure 3) and their oxidation products were analyzed (Figure 4).

The findings shown in Figure 3 indicate that oxidation stability of hydrocarbons decreases with the carbon chain length. Short-chained saturated hydrocarbons can be stored for longer periods of time because of their lower reactivity.

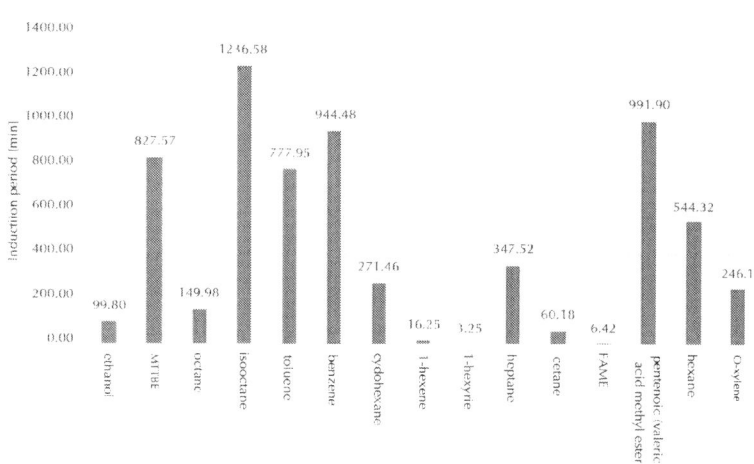

Figure 2: Results of oxidation stability tests of selected chemical compounds.

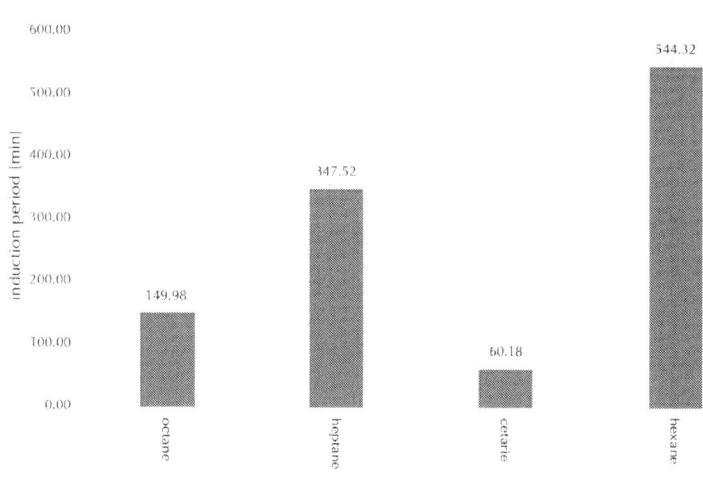

Figure 3: Carbon chain length vs. oxidation stability of hydrocarbons.

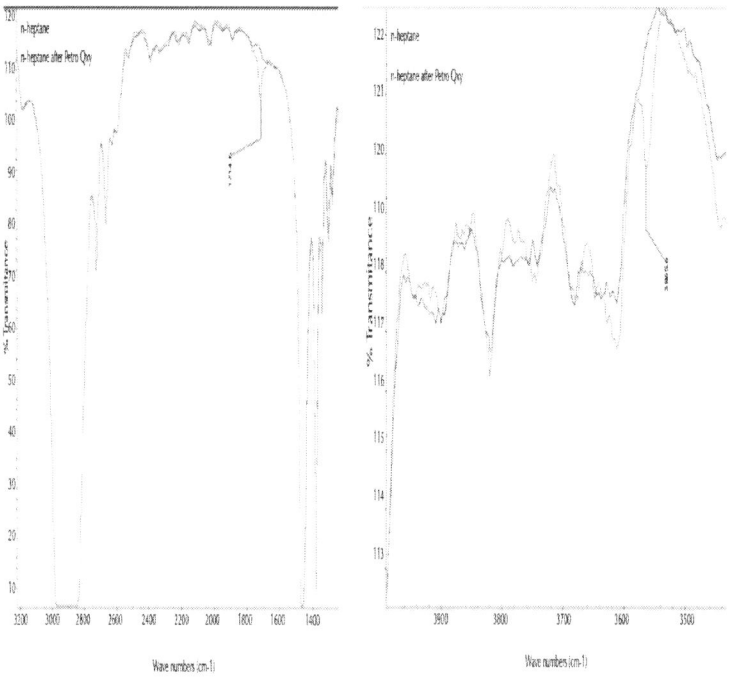

Figure 4: IR spectra for n-heptane before and after ageing (blue-spectrum before ageing, red-spectrum after ageing).

An analysis of the IR spectra of oxidized hydrocarbons indicates that the most significant changes in the n-heptane sample occurred in the wave numbers, corresponding to stretching vibrations of the OH group (approx. 3550 cm^{-1}) and C=O (approx. 1900 cm^{-1}).

From a chromatographic analysis (GC-MS), it follows that the principal products of ageing of heptane are: ketones (i.e., 2-heptanone, 3-heptanone, 4-heptanone) and secondary alcohols: 2-heptanol, 3-heptanol.

Similar changes caused by oxidation were observed in the case of oxidation of n-hexane and n-octane, although the number of carbon atoms per molecule for ketones and alcohols was six and eight, respectively. A different phenomenon was observed for cetane, where the oxidation processes had led to the formation of oxygen compounds (alcohols and ketones) with lower numbers of carbon atoms per molecule, i.e., 2-heptanol, 3-heptanol, 2-heptanone, 3-heptanone, 4-heptanone, 6-dodecanone.

Type of Bonds vs. Oxidation Stability

The type of bonds, especially the unsaturation ratio for a chemical compound or mixture is essential to the oxidation stability of a final product. The more saturated compounds (with single bonds) are present, the better the stability of the mixture. On the contrary, if a double or triple bond is present in a molecule, then oxidation stability decreases dramatically (Figure 5).

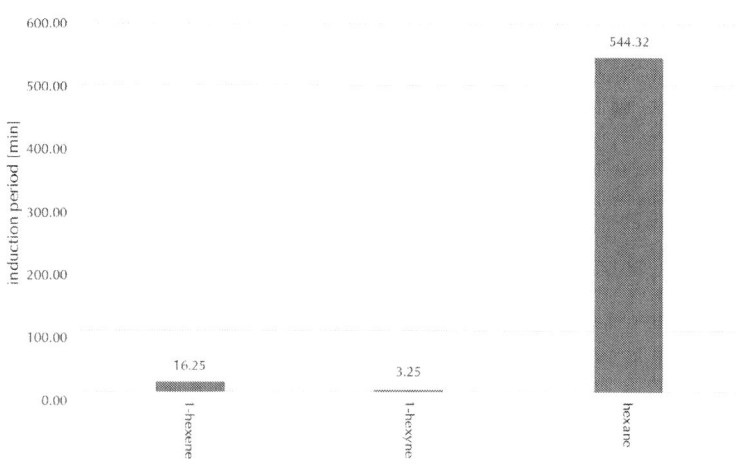

Figure 5: Type of bonds vs. oxidation stability of compounds.

Induction period for hexane is nearly 200 times as high as that for 1-hexyne and more than 30 times as high as that for its double-bond equivalent. Analyses by GC MS and IR indicate that products of oxidation do have a content of aldehydes, carboxylic acids, ketones, and alcohols and aldehydes with a double bond.

Structure vs. Oxidation Stability

Hydrocarbons containing 6 carbon atoms per molecule and having different structures were selected for the structural examination of hydrocarbons on their oxidation stability: hexane, cyclohexane, and benzene.

The oxidation stability tests results are shown in Figure 6.

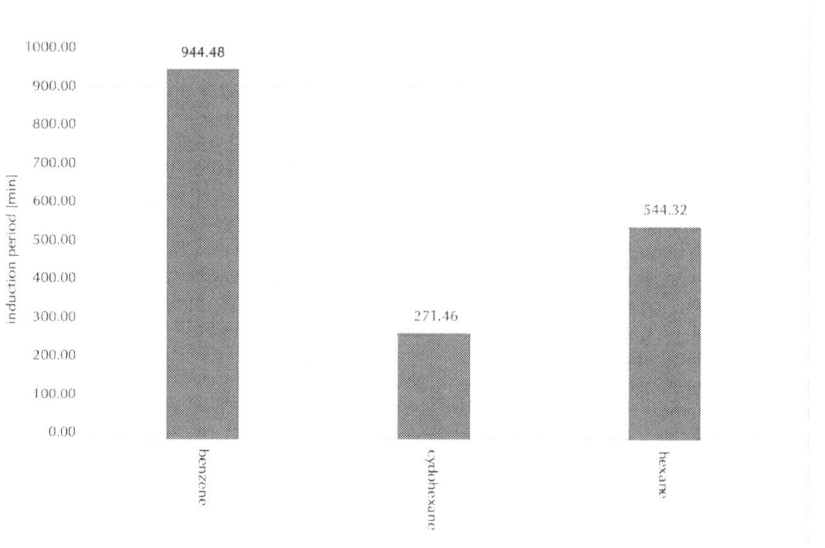

Figure 6: Structure vs. oxidation stability.

The findings indicate that the highest oxidation stability in the above hydrocarbon group was shown by aromatic hydrocarbons with very high chemical stability, followed by hexane, while that of cyclohexane was the lowest. No changes in the IR spectrum of benzene were observed and no new chemical compounds were found after oxidation, as confirmed by the GC MS analysis. Oxidation products included benzaldehyde in the case of toluene, while cyclohexanone and cyclohexanol were found in cyclohexane after oxidation.

The Effect of Isomerization on Oxidation Stability

Chemical compounds with different spatial structures and branching ratios were selected for fuel tests, intended to establish the effect of their isomerization on oxidation stability. The findings shown inFigure 7 indicate that linear compounds are characterized by lower stabilities, compared with their isomers. Induction period for isooctane was nearly 10 times as high as that for octane. This leads us to the conclusion that the oxygen molecule will have a more difficult access to the spatial molecule of isooctane and that the compound will show higher stabilities. In the case of a linear octane molecule, induction period was about 150 minutes. Oxidation of isooctane led to the formation of branched ketones and branched secondary and tertiary alcohols. Oxidation of octane led to the formation of straight-chained ketones and alcohols, such as 2-octanone, 3-octanone, 4-octanone, 2-octanol, 3-octanol, 4-octanol.

Oxidation Stability of Oxygen Compounds

Ethanol and methyl *tert*-butyl ether (MTBE) – components of commercially available motor gasoline – were selected for the tests. The compounds were oxidized by rapid oxidation of which the findings are shown in Figure 8.

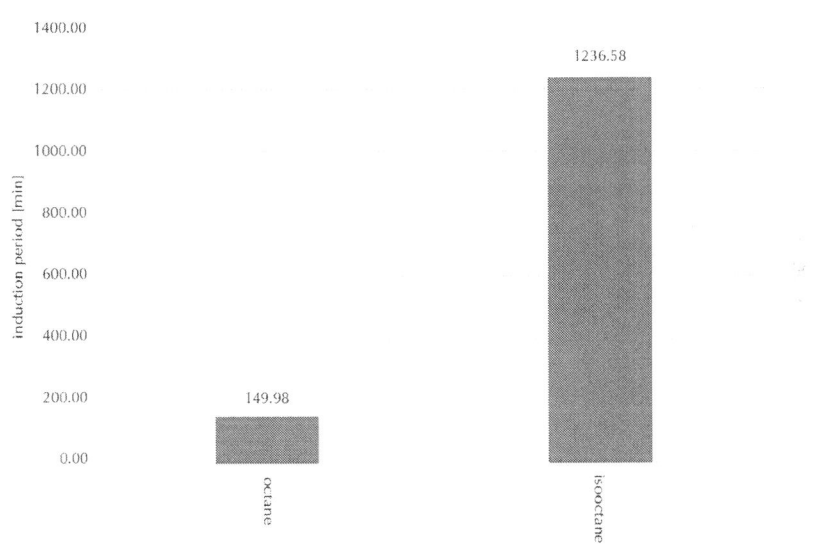

Figure 7: Isomerization vs. oxidation stability of hydrocarbons.

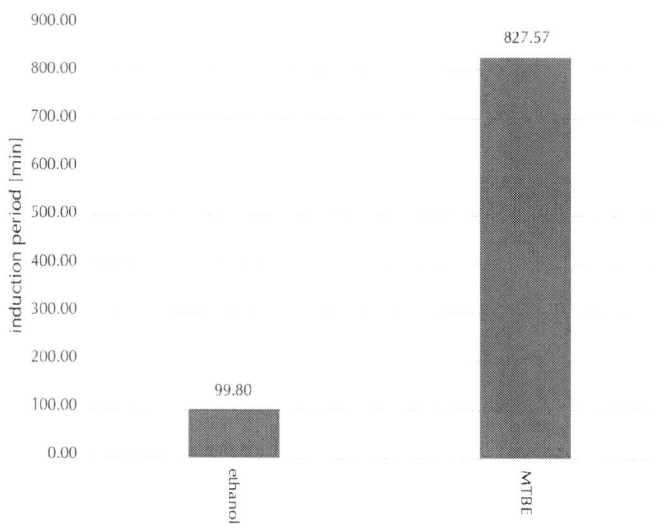

Figure 8: Results of oxidation stability tests for selected oxygen compounds.

The results shown in Figure 8 indicate that ethanol has an induction period nearly 10 times as low as that of methyl *tert*-butyl ether (MTBE). According to the standard EN 228, the maximum permissible content of alcohol in gasoline is 5% (V/V), and the content of ethers is such that the total concentration of oxygen is a maximum of 2.7% (m/m).

AGEING OF MODEL FUEL BLENDS

Stability of Mixture vs. Stability of Compounds

Compounds having extreme oxidation stabilities were mixed together. The experiment was intended to investigate the possibility of addition of oxidation stabilities for the respective compounds and to show the effect of the compound stabilities on that of their mixture.

In the first experiment, two compounds with different oxidation stabilities were mixed together. The resulting mixture was characterized by an oxidation stability which was nearly an arithmetic mean of the two component oxidation stabilities.

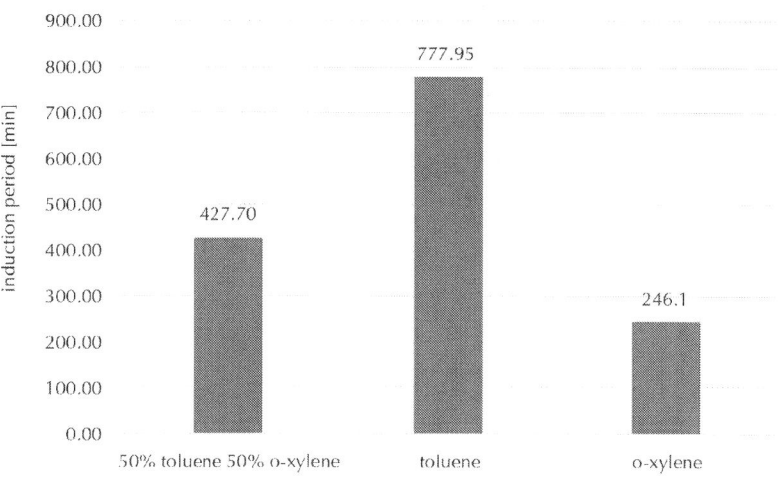

Figure 9: Oxidation stabilities for toluene and o-xylene and their 50/50 mixture.

The other experiments indicate that oxidation stability for the 50/50 mixture is between the two values for the pure individual compounds, though not an arithmetic mean, in a majority of cases.

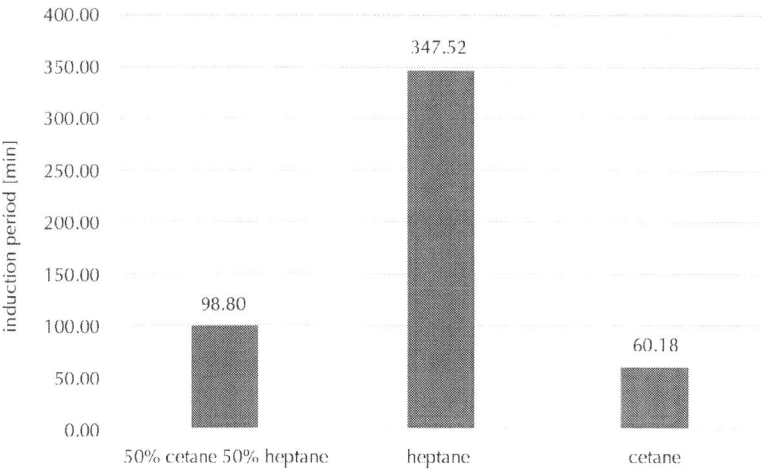

Figure 10: Oxidation stabilities for heptane and cetane and their 50/50 mixture.

Figure 11 indicates that oxidation stability for a mixture of two compounds having extreme values of the parameter is not a mean value of the parameter; rather, the value for the mixture is determined by that of the compound for which the oxidation stability is lower.

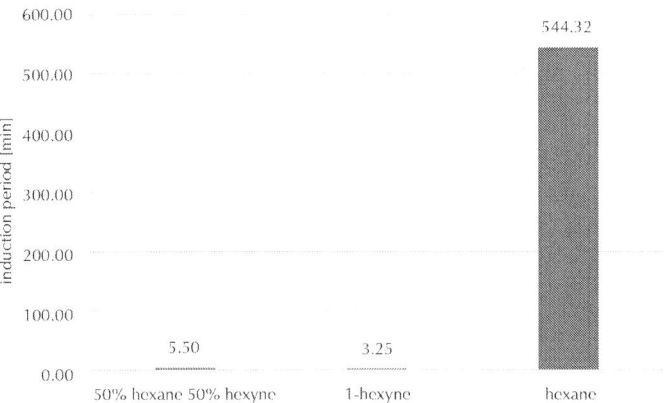

Figure 11: Oxidation stabilities for hexane and 1-hexyne and their 50/50 mixture.

After obtaining a 50/50 mixture of MTBE and isooctane, the value of oxidation stability for the mixture is lower than that for each of the respective chemical compounds.

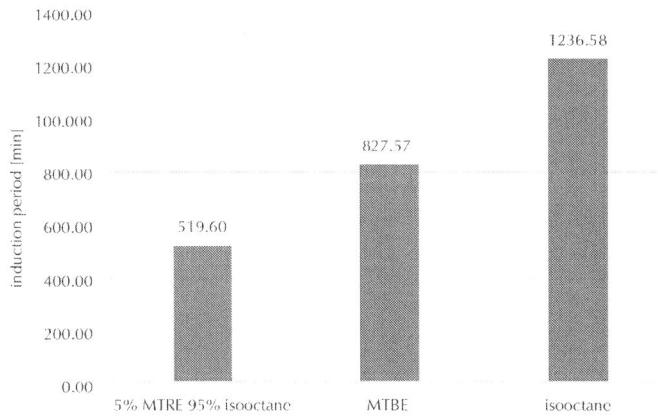

Figure 12: Oxidation stabilities for isooctane and MTBE and their 50/50 mixture.

Oxidation Stability Tests for Fuel Blends

In this experiment, model fuels were prepared and were subjected to ageing by the rapid oxidation method. An analysis was then carried out for samples collected before and after ageing, using the infrared spectrometry technique. Blending was intended to reproduce the actual fuels (gasoline and diesel fuel) as best as possible. In the case of gasoline, blends applicable in octane number determination were used; the octane numbers were approximately the same as that for the gasoline stored for the purposes of this project. In the case of diesel fuel, the mixture was blended by adding to cetane (n-hexadecane) about 20% toluene, just as for actual diesel fuels, in which the content of monocyclic aromatic hydrocarbons is about 20%.

The oxidation stability test results for the model fuel blends are shown in Table 2 and in Figures 13, 14, 15 and 16.

Table 2: Induction period results for the test mixtures

Mixture	Induction period [min]
95% (94 % isooctane + 6 % n-heptane) + 5 % MTBE	670
95% (94 % isooctane + 6 % n-heptane) + 5 % ethanol	519
50% (95 % isooctane + 5 % n-heptane) + 5 % ethanol + 5 % MTBE + 15% 1-heptyne + 25% toluene	183
80 % cetane + 20 % toluene	138.2

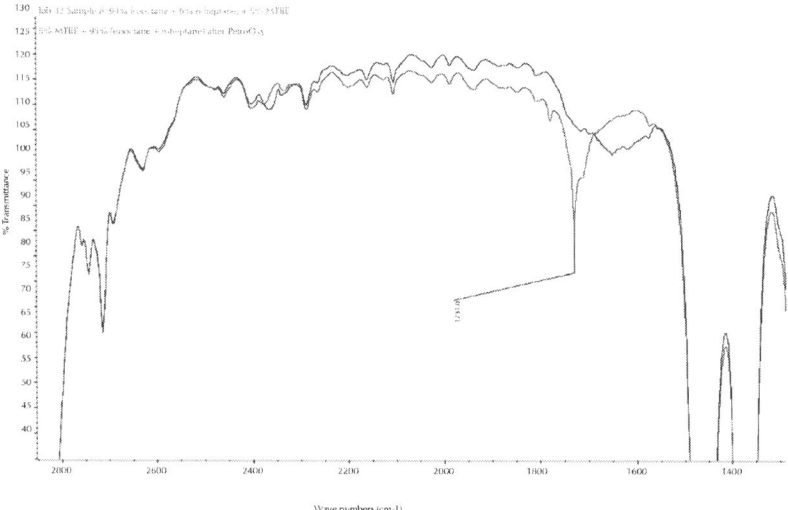

Figure 13: Superimposed IR spectra for 95% (94% isooctane + 6% n-heptane) + 5% MTBE before and after ageing (blue-spectrum before ageing, red-spectrum after ageing).

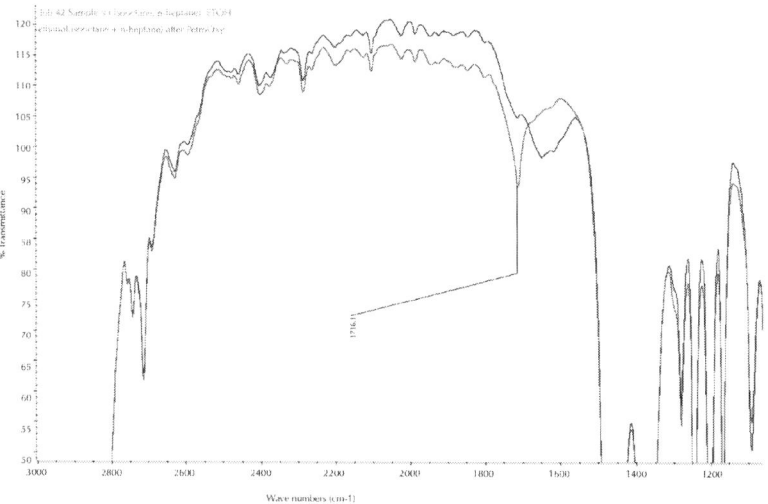

Figure 14: Superimposed IR spectra for 95% (94% isooctane + 6% n-heptane) + 5% ethanol before and after ageing (blue-spectrum before ageing, red-spectrum after ageing).

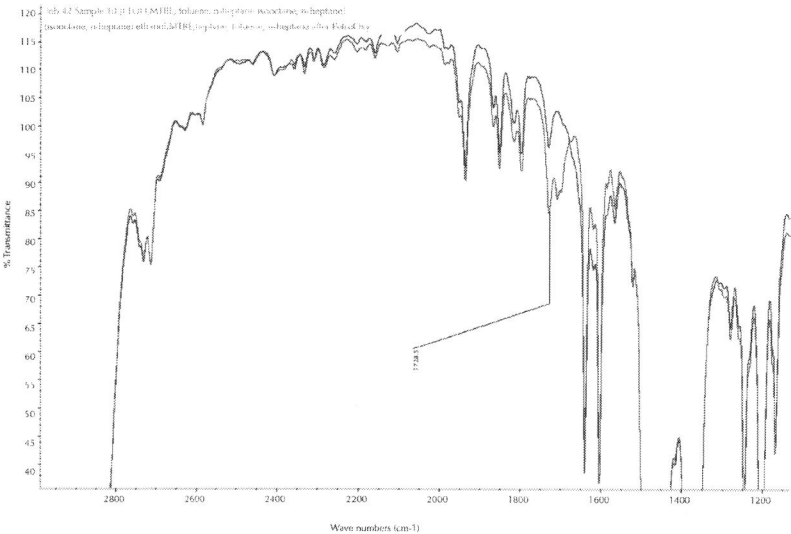

Figure 15: Superimposed IR spectra for 50% (95% isooctane + 5% n-heptane) + 5% ethanol + 5% MTBE + 15% 1-heptyne + 25% toluene before and after ageing (blue-spectrum before ageing, red-spectrum after ageing).

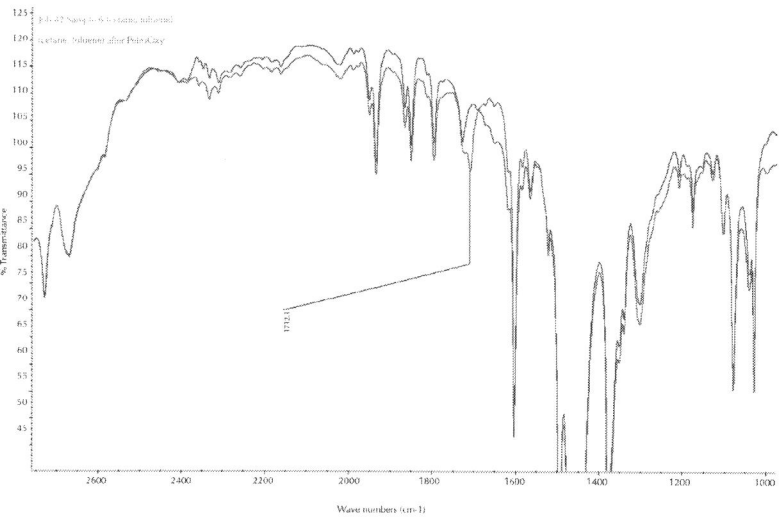

Figure 16: Superimposed IR spectra for 80% cetane + 20% toluene before and after ageing (blue-spectrum before ageing, red-spectrum after ageing).

After oxidation of various types of model mixtures showing fuel compositions after spectral analysis before and after ageing by the rapid-ageing method, all the spectra after oxidation show a characteristic band for the wave numbers in the range 1760-1690 cm^{-1}, corresponding to stretching vibrations from the carbonyl group (C=O). The appearance of the carbonyl group indicates that the sample has been oxidized which means that its initial properties have been changed.

CHANGES IN OXIDATION STABILITY WHICH OCCUR IN CLASSIC MOTOR FUELS DURING LONG-TERM STORAGE

Examination of the various processes taking place in the course of motor fuel ageing during storage was performed using the following fuels:
- 95 octane unleaded gasoline (Pb95),
- 98 octane unleaded gasoline (Pb98),
- diesel fuel with up to 7% (V/V) of biocomponent (ONH),
- diesel fuel with less than 2% (V/V) of biocomponent (ON).

The Pb95 gasoline had a content of not more than 5% (V/V) of a biocomponent in the form of ethyl alcohol. The Pb98 gasoline was an ether-based gasoline, with an ethyl *tert*-butyl ether (ETBE) of not more than 15% (V/V).

The diesel fuels had a content of generation I biocomponent in the form of fatty acid methyl esters (FAME) obtained from rape-seed oil. The ONH diesel fuel had a content of up to 7% (V/V) of the biocomponent. The ON diesel fuel had a content of less than 2% (V/V) of the biocomponent.

The selected fuels were stored for 12 months in 5-m^3 stationary tanks.

The fuels complied with the requirements of applicable standards in respect of the properties of motor gasolines and diesel fuels (EN 228 and EN 590) as at the day of purchase.

The Fuel Storage Station

The purchased motor fuels were stored in 5-m^3 underground storage tanks. The storage tanks were equipped with a dedicated piping system to enable fuel sampling at various heights from the tank. The sampling pumps did not cause mixing of the stored fuel when sampling. Samples of the motor fuels were collected every 2 weeks, as follows: every 4 weeks from level 2 for a full analysis, and every 4 weeks from all levels for a short analysis.

The fuels were sampled at 20 cm, 90 cm and 160 cm above the tank bottom. A fourth pipe was used for collecting fuel vapor samples from above the liquid.

A diagram of the storage tank is shown in Figure 17, the fuel sampling station is shown in Figure 17.

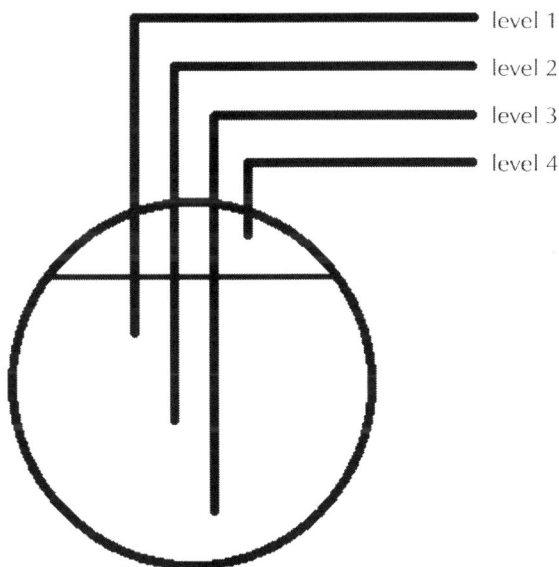

Figure 17: A diagram of the liquid fuel storage tank and the fuel sampling levels in the tank.

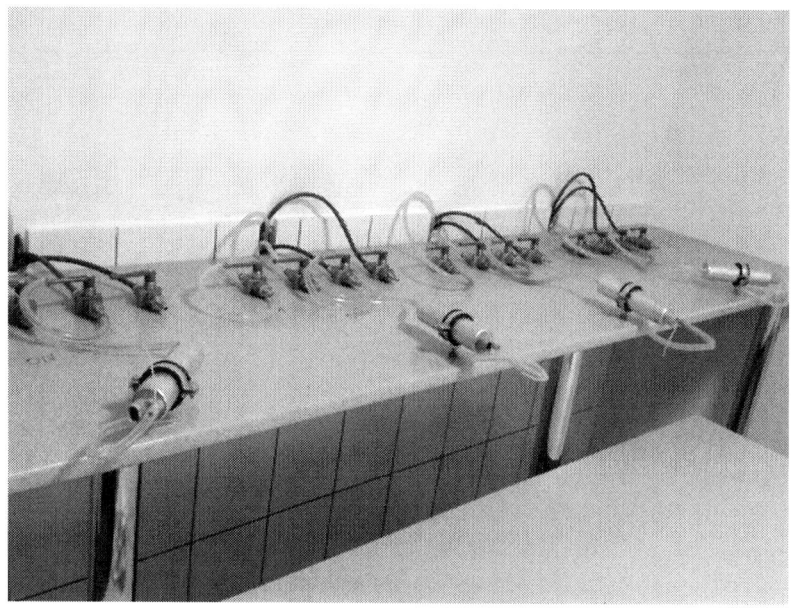

Figure 18: Liquid fuel sampling station.

The scope of analysis was established at the beginning of the fuel storage period.

METHODS OF AGEING OF MOTOR GASOLINES

The following test methods were selected for testing the oxidation stability of motor gasolines:

- oxidation stability test, also referred to as the induction period method according to the standard EN 7536;
- oxidation stability test for small amounts of motor fuels, as described in the standard EN 16091, modified for oxygen pressure (500 kPa).

Induction Period Test

The test method consists in the oxidation with oxygen of the test motor gasoline in a pressurized bomb. A 50 ml volume of the test motor gasoline was placed in a special pressurized bomb (Figure 19) which enables a continuous recording of gas pressure variations, and oxygen was introduced at a pressure in the range 690-705 kPa. The pressurized bomb with the sample in it was then thermostated at 100°C. The test result, referred to as the induction period, is expressed as the time that has lapsed until the maximum pressure has changed by 14 kPa within 15 minutes. The pressure vs. time relationship is plotted in the diagram.

The induction-period method is recommended in the standard EN 228 as the parameter which indicates the quality of the test motor gasoline. A fuel which complies with the standard is expected to have an induction period of more than 360 minutes.

For the induction-period test, samples of the stored motor gasolines were collected from level 2 of the storage tanks (Figure 13) every 4 weeks.

The intention was to continue every test until the gasoline breakpoint. However, the test was discontinued after 60 hrs and it was established that the result was higher than 3600 minutes.

Findings for the test motor gasolines were shown in Figure 20.

The induction-period test results are more than 10 times as high as the minimum value referred to in the standard EN 228, showing very good oxidation stabilities of the stored gasolines.

The selected induction-period method to test motor gasolines in the pressurized bomb did not enable it to be established which of the stored motor gasolines had a better oxidation stability.

Figure 21 shows an oxidation curve for the Pb95 and Pb98 motor gasolines after storing them for 50 weeks, as found in accordance with the standard EN 7536.

Figure 21 indicates that ageing runs faster for the Pb95 gasoline, compared with the Pb98 gasoline. The faster drop of oxidation stability for Pb95 is caused by the addition of ethanol, of which the reactivity with oxygen is higher than that of ether.

94   Chemical Process Equipment: Selection and Design

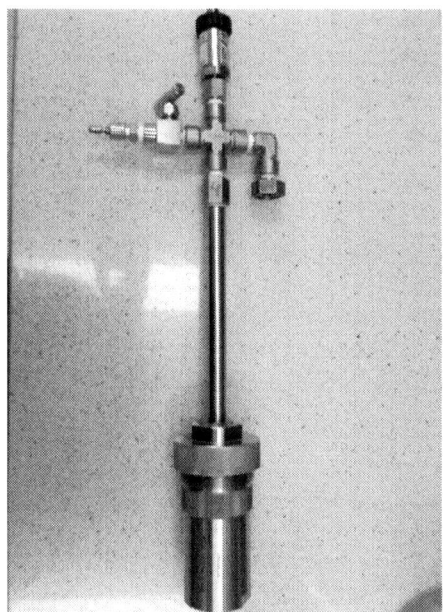

Figure 19: Pressurized bomb for oxidation stability tests of motor gasolines.

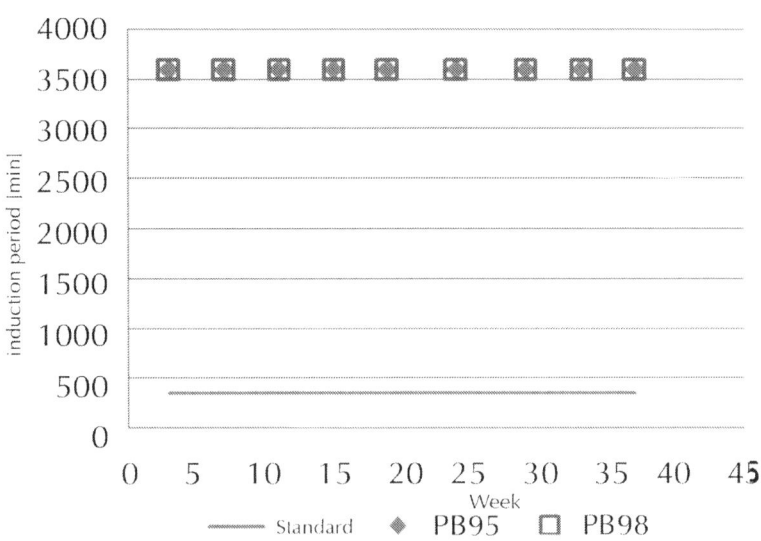

Figure 20: Graphical representation of findings for the test motor gasolines.

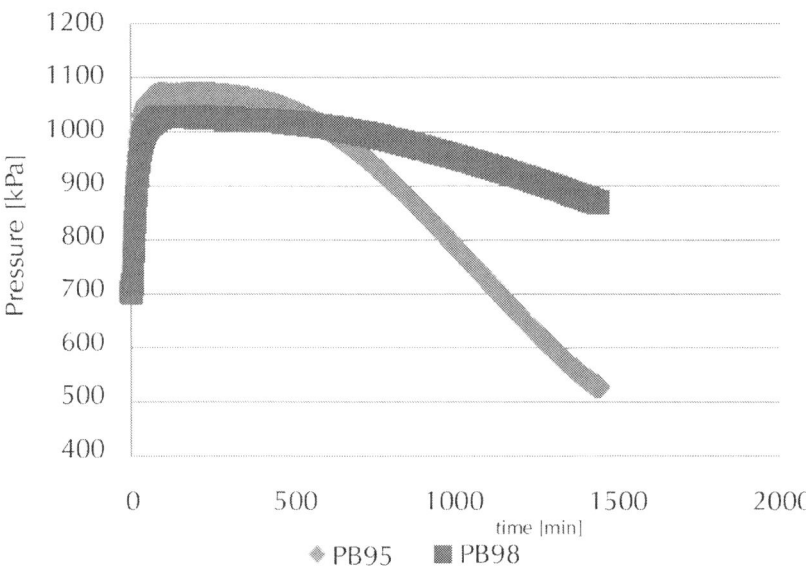

Figure 21: The course of oxidation of motor gasolines after 50 weeks of storage, as found by the induction-test method.

The Rapid-Oxidation Test of Motor Gasolines

The rapid-oxidation test, intended to measure oxidation stability of motor gasolines, was carried out in the apparatus shown in Figure 22. The test method is described in the standard EN 16091. The method has the advantage of requiring only a small volume of the test sample. The oxidation stability test of motor gasolines was carried out at the following conditions: temperature: 140°C, pressure of oxidizing factor 500 kPa, test sample volume: 5 ml. Oxygen was used as the oxidizing factor.

The oxidation stability test of motor gasolines was carried out for stored motor gasolines every 2 weeks, using the rapid-ageing method. Figure 23 shows the test results for the Pb95 and Pb98 gasolines after storage.

Only a small sample volume is required for the test, therefore, the method was used also for testing gasoline samples collected from various tank levels to find out whether fractionation occurred during stationary storage. Findings for the 95-octane and 98-octane gasolines are shown in Figure 24 andFigure 25, respectively.

Figure 26 shows the course of a representative test – the diagram illustrates a change in the pressure generated in the test vessel vs. time for the test fuel sample.

The method confirms earlier observations that the gasoline with a biocomponent content (bioethanol) has a lower oxidation stability, compared with that which contains ethyl *tert*-butyl ether.

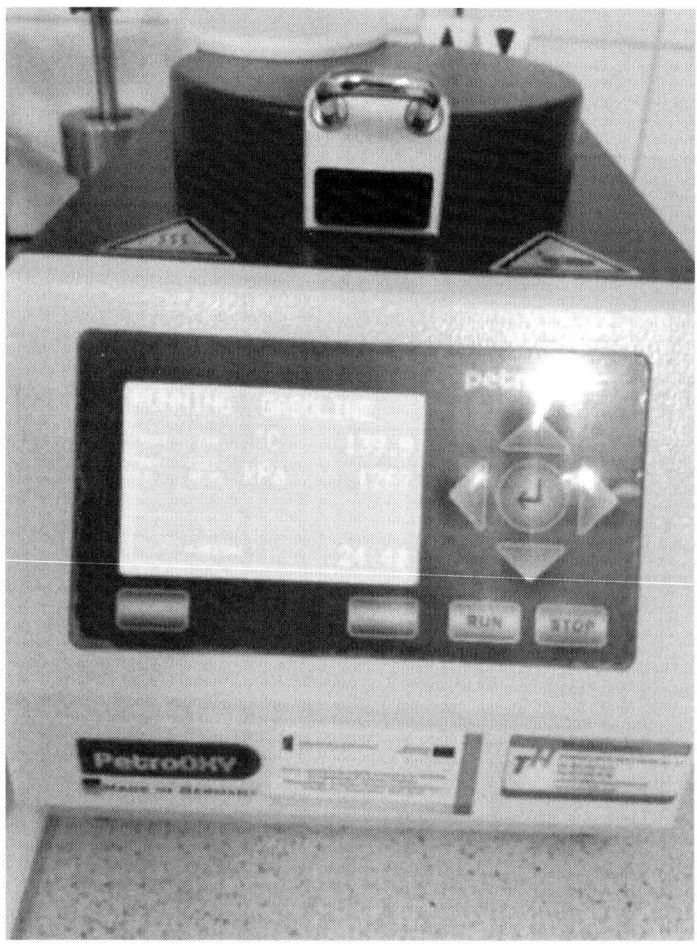

Figure 22: Oxidation stability test apparatus.

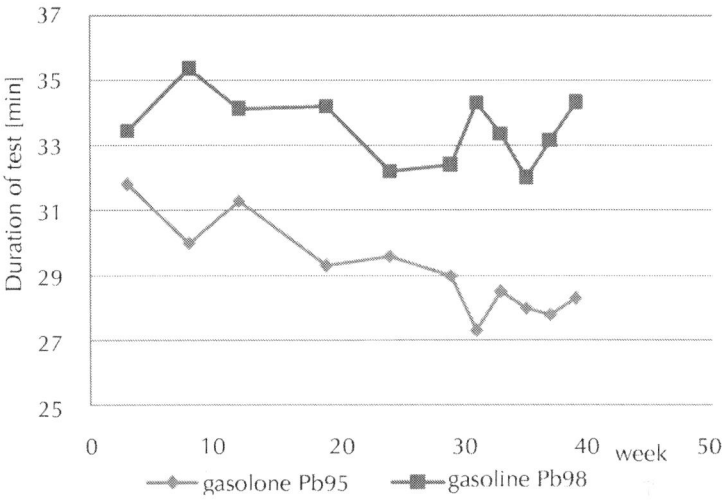

Figure 23: Results of the rapid-oxidation test for PB95 and PB98 gasolines.

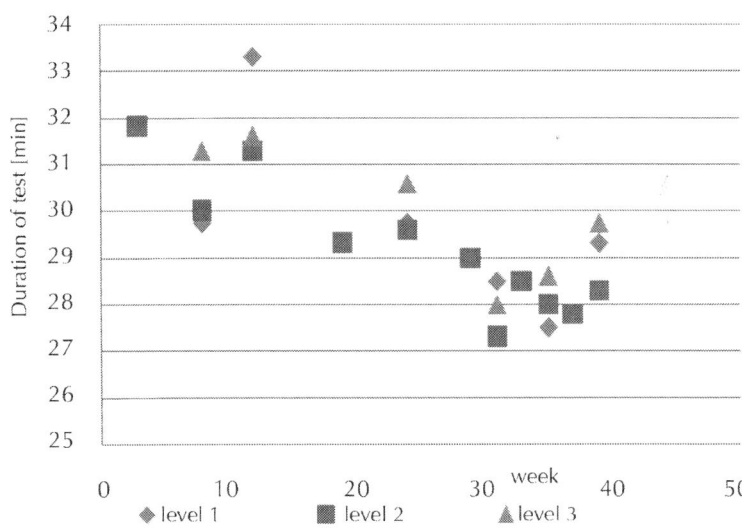

Figure 24: Results of the oxidation stability test for the PB95 gasoline, collected from various tank levels.

98 Chemical Process Equipment: Selection and Design

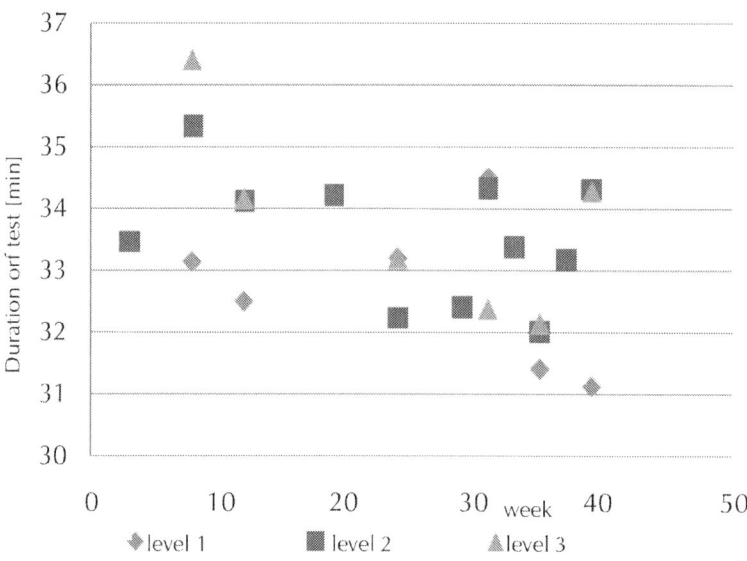

Figure 25: Results of the oxidation stability test for the Pb98 gasoline, collected from various tank levels.

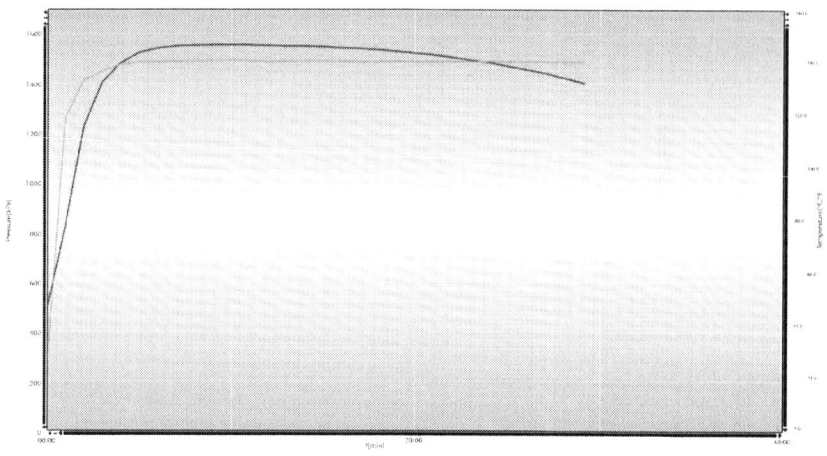

Figure 26: Graphical representation of oxidation of the test motor gasoline.

DIESEL FUEL AGEING METHODS

Oxidation stability tests were carried out on diesel fuels after storage, using the following methods:
- determination of the amount of deposit formed in diesel fuels during oxidation;
- determination of oxidation stability of diesel fuels having a content of biocomponents;
- rapid-oxidation test for diesel fuels.

Determination of Oxidation Stability from the Amount of Deposit Formed

The method consists in oxygen-based oxidation of diesel fuel test samples at a flow rate of 3 l/hr and a temperature of 95°C. The test is continued for 16 hours, during which time reactions take place in the diesel fuel sample, leading to the formation of macromolecular organic compounds. The quality of diesel fuel is evaluated from the amount of deposit formed as the result of oxidation of the fuel. According to the standard EN 590, a fuel meets the requirement if the amount of deposit is less than 25 g/m^3. The method is dedicated to all types of diesel fuel and is carried out in accordance with the methodology, described in the standard EN 12205. Figure 27 shows the components of the test assembly for diesel fuel ageing (thermostating bath for the test sample, test tube with a suitable condenser and a filtering assembly).

100   Chemical Process Equipment: Selection and Design

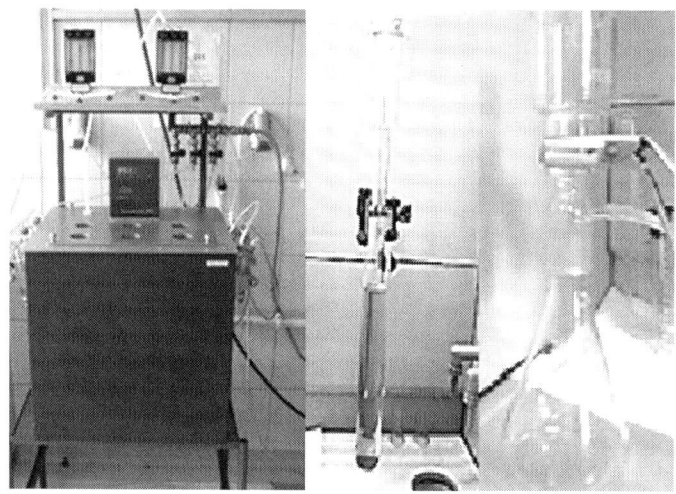

Figure 27: A test assembly for oxidation-stability tests of diesel fuels according to the standard EN 12205.

Oxidation stability was tested using the method for two types of diesel fuel after storage: one of them (ON) had a content of biocomponents of less than 2% (V/V), the other was a diesel fuel with 7% (V/V) of a biocomponent (ONH). In this method, oxidation stability is evaluated from the sum of adherent solubles and filterable insolubles, present in the diesel fuel after oxidation.

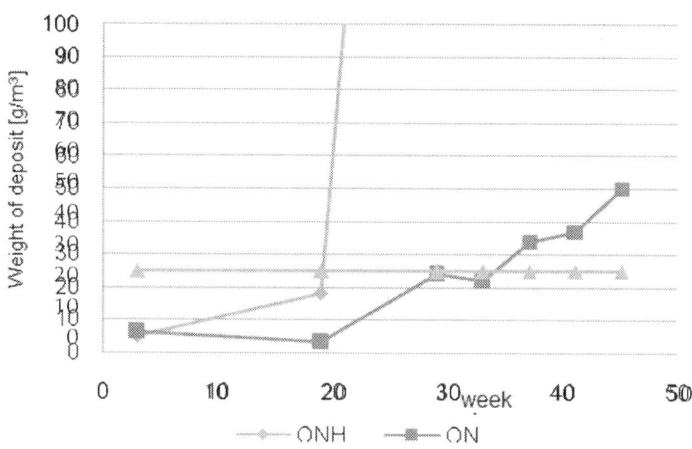

Figure 28: Results of oxidation stability tests.

Figure 28 is a graphical representation of the oxidation-stability test results, obtained by the test method described in the standard EN 12205. The blue color in the diagram shows results for the diesel fuel with 7% (V/V) of a biocomponent, orange color shows the results for the diesel fuel with less than 2% (V/V) of a biocomponent. The grey line shows the permissible value of oxidation stability for diesel fuels (25 g/m^3) according to the standard EN 590.

The results in Figure 28 indicate that the content of deposit in the diesel fuel with 7% (V/V) of a FAME after being stored for 20 weeks is so high that the diesel fuel becomes virtually unsuitable for use as a fuel for combustion engines. In Week 45 of storage, the diesel fuel had a content of solubles (gums) of more than 3000 g/m^3. The solubles and insolubles formed are shown in Figures 29 and 30.

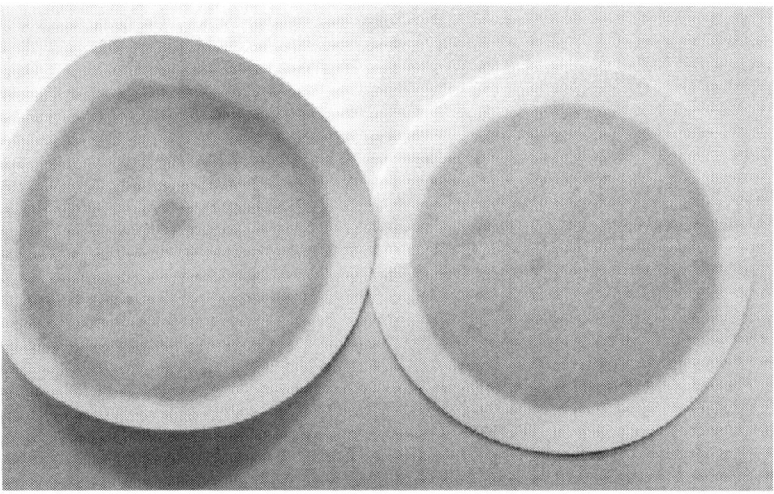

Figure 29: Insolubles, formed by oxidation during storage of diesel fuel with 7% (V/V) of a biocomponent.

Figure 30: Solubles (gums), formed by oxidation during storage of diesel fuel with 7% (V/V) of a biocomponent.

For the diesel fuel with less than 2% (V/V) of a FAME, ageing is a much slower process. In Week 33 of storage, the deposit content in the diesel fuel is near the maximum value of 25 g/m^3, as stated in the standard EN 590.

The results of oxidation stability tests, carried out by the method referred to in the standard EN 12205, indicate that the diesel fuel with 7% (V/V) of a FAME is affected by ageing at a faster rate, compared with that containing less than 2% (V/V) of a biocomponent.

Oxidation Stability Tests of Diesel Fuel with More than 2% (V/V) of Biocomponents (Rancimat Method)

Oxidation stability tests of diesel fuels are carried out in accordance with EN 15751 and dedicated, according to the standard EN 590, only to fuels with more than 2% (V/V) biocomponents in the form of fatty acid methyl esters (FAME).

The test method is based on oxidation under the effect of air (10 l/hr) and temperature (110°C), causing the ester to decompose, whereby acid compounds are released which affect the electrical conductivity of water, as controlled during the test.

The test is carried out at a temperature of 110°C and air flow rate of 10 l/hr. The required sample volume is 7.5 g of product. Diesel fuels comply with the requirements of the standard EN 590, if the time that lapses until the break point in the electrical conductivity curve exceeds 20 hrs.

The test was carried out using the device shown in Figure 31.

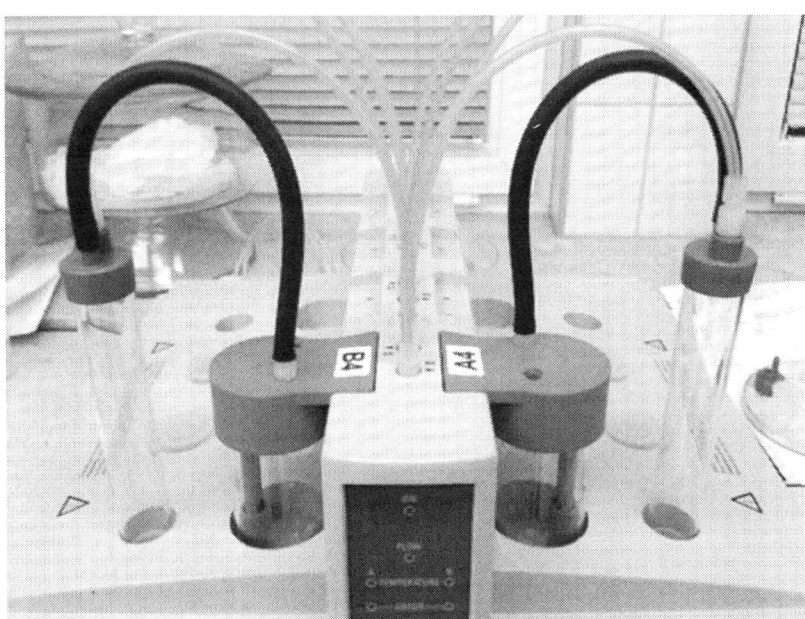

Figure 31: A test assembly for oxidation stability tests according to EN 15751 (Rancimat method).

The test result, expressed in hours, is found from the curve which illustrates changes in the electrical conductivity of water vs. time (Figure 32).

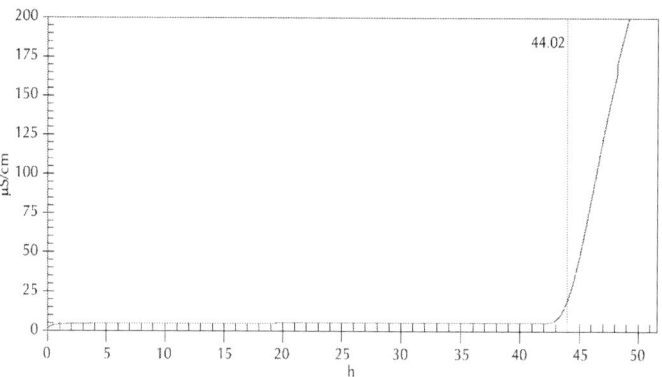

Figure 32: The course of oxidation of diesel fuel according to EN 15751 (Rancimat method).

The tests results were shown in Figure 33. The blue line shows the ageing processes taking place in the diesel fuel with 7% (V/V) of a FAME, the red line refers to the oil with less than 2% (V/V) of a biocomponent. The oil with less than 2% (V/V) of a biocomponent was tested for comparison of its results with those obtained for the oil with 7% (V/V) of FAME.

The results seem to confirm the observation that ageing processes run at a faster rate in the diesel fuel with 7% (V/V) of a FAME, compared with the oil containing less than 2% (V/V) of a biocomponent.

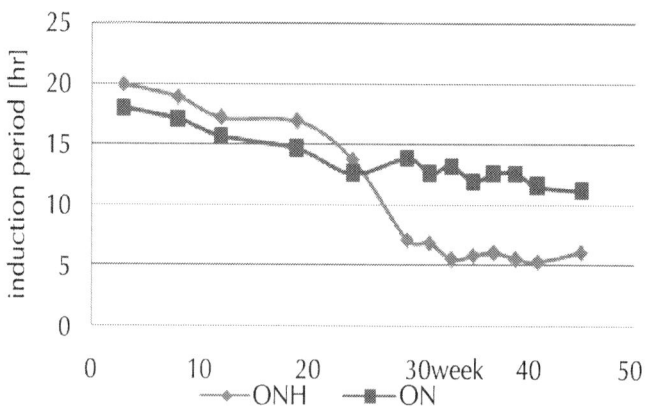

Figure 33: Results of ageing for diesel fuels during storage.

Rapid-Oxidation Method for Diesel Fuels

The third method to test the diesel fuels for oxidation stability is the most recent method, recommended for the purpose. The test procedure was described in the standard EN 16091. The criterion of evaluation of the quality of diesel fuels is the time that has lapsed until pressure in the reaction vessel has dropped by 10%, compared with its initial value, and is expressed in minutes. The test method has the advantage of short duration, requiring only a small sample volume and providing results with high repeatability.

The procedure and conditions of the oxidation stability test were the same as for motor gasolines (Item 6.2.), except that the initial pressure of oxygen was 700 kPa for diesel fuel.

The fuel samples were collected at level 2 of the fuel tank every 2 weeks.

The oxidation stability test results for the diesel fuels during storage are shown in Figure 34. The blue line refers to the diesel fuel with 7% (V/V) of a biocomponent (ONH), the red line indicates the fuel with less than 2% (V/V) of a FAME biocomponent (ON).

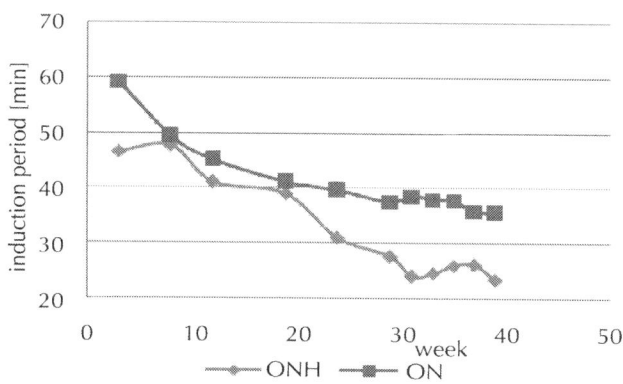

Figure 34: Results of oxidation stability tests for diesel fuels during storage.

The results of oxidation stability tests, as shown in the diagram, indicate that the diesel fuel with 7% (V/V) of FAME is much inferior in respect of oxidation stability, compared with the diesel fuel with less than 2% (V/V) of FAME.

While the ONH diesel fuel sample, after being stored for 40 weeks in an underground storage tank, had an induction period of 23 minutes, the ON diesel fuel sample had an induction period more than 50% higher, indicating a superior oxidation stability of the diesel fuel with the lower content of FAME biocomponent.

CONCLUSIONS

Based on oxidation stability tests, an increase in the chain length of hydrocarbons was found to cause a decrease in oxidation stability. The longer the hydrocarbon chain of a compound, the lower its stability. The presence of multiple bonds leads to lower oxidation stabilities. Unsaturated compounds display higher reactivities, compared with saturated ones: they tend to be oxidized and polymerized more readily. Cyclic compounds are less stable than aromatic or aliphatic compounds because of the presence of stresses in their ring. Aromatic compounds are more stable because of the presence of a system of three coupled double bonds in their molecules.

Toluene has a lower oxidation stability compared with benzene because aromatic compounds with a hydrocarbon chain are more susceptible to oxidation compared with rings without substitutes. The aromatic ring is highly stable and double bonds will resist cleavage. If, therefore, a single bond exists between a carbon atom in the ring and a carbon atom in the substitute, then location at the substitute's carbon atom is preferential for the incorporation of oxygen atoms (oxidation of toluene produces benzaldehyde).

Branched isomers are more stable than linear hydrocarbons because interactions between carbons atoms in the branched isomers are more difficult.

Ethyl alcohol is oxidized faster than methyl *tert*-butyl ether (MTBE) because alcohols tend to react with oxygen more readily than ethers. Oxidation of ethers is a slow reaction: a GC MS analysis showed the presence of methyl *tert*-butyl ester after contact with oxygen.

Methyl esters of saturated acids are more stable than methyl esters of unsaturated acids (FAME). This is caused by the presence of unstable multiple bonds.

Oxidation of saturated compounds produces mainly alcohols and ketones. The reactions run at a slow rate, although the presence of such compounds is indicative of oxidation processes taking place.

Oxidation of unsaturated compounds leads to the formation of alcohols, ketones, aldehydes, and carboxylic acids.

Ethyl alcohol and methyl *tert*-butyl ether (MTBE) are oxidized to esters, although the reactions run at different rates.

The appearance of the carbonyl group as the result of accelerated oxidation of mixtures designed to imitate the composition of fuels, confirms the fact that oxidation has taken place, thereby, its initial properties have changed.

Motor gasolines with a content of ethanol have a lower oxidation stability, compared with those having a content of ether compounds. This is also confirmed by the fact that alcohols are oxidized more readily than ethers.

A diesel fuel with 7% (V/V) of FAME has a lower oxidation stability, compared with the diesel fuel containing less than 2% (V/V) of FAME; consequently, fatty acid methyl esters determine the stability of final products because of their different chemical structure (multiple bonds), among other things. The fact is confirmed by all oxidation stability analyses, including the Rancimat method, PetroOxy, and the weighing method to determine deposits formed by sample oxidation.

REFERENCES

1. Ed.: Jan Surygała, Vademecum rafinera [A refiner's guide], Wydawnictwo Naukowo-Techniczne, Warsaw, 2006
2. Baczewski K., Kałdoński T. Paliwa do silników o zapłonie samoczynnym [Fuels for self-ignition engines], Wydawnictwa Komunikacji i Łączności, 2004, 2008
3. Baczewski K., Kałdoński T. Paliwa do silników o zapłonie iskrowym [Fuels for spark-ignition engines], Wydawnictwa Komunikacji i Łączności, 2005,
4. Witkiewicz Z. Podstawy chromatografii [Foundations of chromatography], Wydawnictwo Naukowo-Techniczne, Warsaw

5. Kęcki Z. Podstawy spektroskopii molekularnej [Foundations of molecular spectroscopy](Pages 15-26, 56-80, 84-91, 143-155)

Chapter 5

Optimal Solutions of Multiproduct Batch Chemical Process Using Multiobjective Genetic Algorithm with Expert Decision System

Diab Mokeddem and Abdelhafid Khellaf

Department of Electronics, Faculty of Engineering, University of Setif, 19000 Setif, Algeria

ABSTRACT

Optimal design problem are widely known by their multiple performance measures that are often competing with each other. In this paper, an optimal multiproduct batch chemical plant design is presented. The design is firstly formulated as a multiobjective optimization problem, to be solved using the well suited non dominating sorting genetic algorithm (NSGA-II). The NSGA-II have capability to achieve fine tuning of variables in determining a set of non dominating solutions distributed along the Pareto front in a single run of the algorithm. The NSGA-II ability to identify a set of optimal solutions provides the decision-maker DM with a complete picture of the optimal solution space to gain better and appropriate choices. Then an outranking with PROMETHEE II helps the decision-maker to finalize the selection of a best compromise. The effectiveness of NSGA-II method with multiojective optimization problem is illustrated through two carefully referenced examples.

INTRODUCTION

Batch processes are used in production of many low-volume but high-value-added products (such as speciality chemicals, health care, food, agrochemicals,…etc.) because of operation flexibility in today's market-driven environment. Manufactory of these products generally involves multi step synthesis [1]. In addition, if two or more products require similar processing steps, the same set of equipment is considered for at least economical reason. A batch plant producing multiple products is categorized as either a multiproduct plant or a multipurpose plant. Multiproduct plants produce multiple products following a sequential similar recipe. In such a plant, all the products follow the same path through the process and only one product is manufactured at a time. Each step is carried out on single equipment or on several parallel equipment units. Processing of other products is carried out using the same equipment in successive production runs or campaigns. In a multipurpose plant, each product follows one or more distinct processing paths; so more than one product may be produced simultaneously in such plants. The present work is directed toward the optimal design problems of multiproduct batch plants.

In conventional optimal design of a multiproduct plant, production requirements of each product and a total production time for all products are available and specified. The number, the required volume, and size of parallel equipment units in each stage are then determined to minimize the investment. It should be emphasized that batch plantsdesign has been for long identified as a key problem in chemical engineering as reported in literature [2–9]. Formulation of batch plant design generally involves mathematical programming methods, such as linear programming (LP), nonlinear programming (NLP), mixed-integer linear programming (MILP) or mixed-integer nonlinear programming (MINLP). Mathematical programming or different optimization techniques, such as branch and bound, heuristics, genetic algorithm, simulated annealing, are thoroughly used to derive optimal solutions.

However, in reality the multiproduct design problem can be formulated as a multiobjective design optimization problem in which one seeks to minimize investment, operation cost, and total production time, and, simultaneously, to maximize the revenue. Recall that not much work has been reported in the literature on the multiobjective optimal design of a multiproduct batch plant. Huang and Wang [10] introduced a fuzzy decision-making approach for multiobjective optimal design problem of a multiproduct batch plant. A monotonic increasing or decreasing membership function is used to define the degree of satisfaction for each objective function and the problem is then represented as an augmented minmax problem formulated as MINLP models. To obtain a unique solution, the MINLP problem is solved using a hybrid differential evolution technique. Dedieu et al. [11] presented the development of a two-stage methodology for multiobjective batch plant design and retrofit according to multiple criteria. The authors used a multiobjective genetic algorithm based on the combination of a single-objective genetic algorithm and a Pareto sort procedure for proposing several plant structures and a discrete event simulator for evaluating the technical feasibility of the proposed configurations.

In the case of multiple objectives, an optimum solution with respect to all objectives may not exist. In most cases, the objective functions are in conflict, because in order to decrease any of the objective functions, we need to increase other objective functions. Recently, Solimanpur et al. [12] developed a sophisticated multiobjective integer programming model where the objectives considered were the maximization of total similarity between parts, the minimization of the total processing cost, the minimization of the total processing time and the minimization of the total investment needed for the acquisition of machines [13].

The presence of multiple objectives in a problem usually gives rise to a family of nondominated solutions, largely known as Pareto-optimal solutions, where each objective component of any solution along the Pareto front can only be improved by degrading at least one of its other objective components. Since none of the solutions in the nondominated set is absolutely better than any other, any one of them is then an acceptable solution. As it is difficult to choose any particular solution for a multiobjective optimization problem without iterative interaction with the decision maker (DM) [14] one general approach is to establish first the entire set of Pareto-optimal solutions, where an external Decision Maker (DM) direct intervention gives interactive information in the multiobjective optimization loop [15]. So, a satisfactory solution of the problem is found as soon as the knowledge is acquired [16]. Promethee II (Preference Ranking Organisation METHod for Enrichment Evaluations—2nd version) is a popular decision method that has been successfully applied in the selection of the final solution of multiobjective optimization problems. It generates a ranking of available points, according to the DM preferences, and the best ranked one is considered the favourite final solution. It is based on the concept of outranking relation, which is a binary relation defined between every pair (a,b) of alternatives, in such way that, if a is preferred to b (according to the DM interests), then it is said that $a$outranks b. When these relations are defined between all pairs of alternatives, they are exploited according to some rules in order to rank all solutions from the best to the worst.

The first GA proposed for multiobjective optimization was VEGA [17]. This is a nonPareto based approach based on the selection of several relevant groups of individuals, each group being associated to a given objective. It is reported that the method tends to crowd results at extremes of the solution space, often yielding to poor convergence of the Pareto front. A more recent algorithm, based on scalarization with a weighted sum function, is proposed in Ishibuchi and Murata [18] where the weights are randomly chosen. Many successful evolutionary multiobjective optimization algorithms were developed based on the two ideas suggested by Goldberg [19]: Pareto dominance and niching. Pareto dominance is used to exploit the search space in the direction of the Pareto front and niching technique explores the search space along the front to keep diversity. The well-known algorithms in this category include Multiobjective Genetic Algorithm: (MOGA) [16], Niched Pareto Genetic Algorithm: (NPGA) [20], Strength Pareto Evolutionary Algorithm: (SPEA) [21], Multiobjective Evolutionary Algorithm: (MOEA) [22], the Nondominated Sorting Genetic Algorithm (NSGA) proposed by Srinivas and Deb [23] was one of the first evolutionary algorithm for solving multiobjective optimization problems. Although NSGA has been successfully applied, the main criticisms of this approach has been its high computational complexity of nondominated sorting, lack of elitism, and need for specifying a tuneable parameter called sharing parameter. Recently, Deb et al. [24] reported an improved version of NSGA, which they called NSGA-II, to address all the above issues.

The purpose of this study is to extend this methodology for solution of multiobjective optimal control problems under the framework of NSGA-II. The efficiency of the proposed method is illustrated by solving multiobjective optimization problem.

FORMULATION OF THE MULTIOBJECTIVE PROBLEM

The problem of multiproduct batch plant covered in this paper can be defined by assuming that the plant consists of a sequence of M batch processing stages that are used to manufacture N different products. At each stage j there are N_j identical units in parallel operating out of phase, each with a size V_j. Each product i follows the same general processing sequence.

Batches are transferred from one stage to the next without any delay, that is, we consider a zero-wait operating policy.

In the conventional design of a multiproduct batch plant, one seeks to minimize the investment cost by determining the optimal number, required volume and size of parallel equipment units in each stage for a specified production requirement of each product and the total production time. However, in reality the designer considers not only minimizing the investment but also minimizing the operation cost and total production time while maximizing the revenue, simultaneously

$$\underset{N_j,V_j,B_i,T_{L_i},Q_i,H}{\text{Max}} \quad \text{Revenue} = f_1 = \sum_{i=1}^{N} Cp_i Q_i, \qquad (1)$$

M

$$\underset{N_j,V_j,B_i,T_{L_i},Q_i,H}{\text{Min}} \quad \text{Investment cost} = f_2 = \sum_{j=1}^{M} N_j \alpha_j V_j^{B_j}, \quad (2)$$

$$\underset{N_j,V_j,B_i,T_{L_i},Q_i,H}{\text{Min}} \quad \text{Operation cost} = f_3 = \sum_{i=1}^{N} \sum_{j=1}^{M} C_{E_i} \frac{Q_j}{B_i} + C_{o_i} Q_i,$$

$$(3)$$

$$\underset{N_j,V_j,B_i,T_{L_i},Q_i,H}{\text{Min}} \quad \text{Total production time} = f_4 = H. \quad (4)$$

So, the multiobjective problem consists of determining the following parameters:

- N_j the number of parallel units in stage j,
- V_j the required volume of a unit in stage j,
- B_i size of the batch of product i at the end of the M stages,
- TL_i the cycle time for product i,
- Q_i the production requirement of product i and,
- (H the total production time,

while satisfying certain constraints such as volume, time, and so forth.

The constraints are expressed as follows:

- Volume constraints. Volume Vj has to be able to process all the products i:

$$S_{ij} B_i \leq V_j, \quad \forall i = 1,\ldots,N; \quad \forall j = 1,\ldots,M. \quad (5)$$

- Time constraint. The summation of available production time for all products is not more than the net total time for production

$$\sum_{i=1}^{N} \frac{Q_i}{B_i} T_{L_i} \leq H. \quad (6)$$

- The limiting cycle time for product i:

$$\frac{T_{ij}}{N_j} \leq T_{L_i}, \quad \forall i = 1,\ldots,N; \quad \forall j = 1,\ldots,M. \quad (7)$$

- Dimension constraints. Every unit has restricted allowable range

$$V_j^L \leq V_j \leq V_j^U, \quad \forall j = 1,\ldots,M,$$
$$B_j^L \leq B_j \leq B_j^U, \quad \forall j = 1,\ldots,N. \tag{8}$$

ELITIST NONDOMINATED SORTING GENETIC ALGORITHM (NSGA-II)

The NSGA II Pareto ranking algorithm is an elitist Deb et al. [24] system and maintains an external archive of the Pareto solutions. In contrast to the simple genetic algorithms that look for the unique solution, the multiobjective genetic algorithm tries to find as many elements of the Pareto set as possible. For the case of the NSGA-II, this one is provided with operators who allow it to know the level of nondominance of every solution as well as the grade of closeness with other solutions; which allows it to explore widely inside the feasible region.

In a brief form, the functioning of the multiobjective genetic algorithm NSGA-II can be described through the following steps.

Fast Nondominated Sort

A very efficient procedure, is used to arrange the solutions in fronts (nondominated arranging), in accordance with their aptitude values. This is achieved, creating two entities for each of the solutions. A domination count np, the number of solutions which dominates the solution p, and a set (Sp), that contains the solutions that are dominated for p. The solutions of the first front have the higher status of nondominance in the Pareto sense.

Diversity Preservation

This is achieved, by means of the calculation of the crowding degree or closeness for each of the solutions inside the population. This quantity is obtained, by calculating the average distance of two points on either side of a particular solution along each of the objectives. This quantity serves as an estimate of the cuboid perimeter, formed by using the nearest neighbours as the vertices. There is also, an operator called Crowded-Comparison (≺n), which guides to the genetic algorithm, towards the Pareto optimal front, in accordance with the following criterion:

$$i \prec_n j \text{ if } (i_{rank} < j_{rank}),$$

$$\text{or } (i_{rank} = j_{rank}) \text{ and } (i_{didtance} > j_{distance}).$$

In accordance with the previous criterion, between two nondominated solutions, we prefer the solution with the better rank. Otherwise, if both solutions belong to the same front, then, we prefer the solution that is located in a lesser crowded region.

Initial Loop

Initially, a random parent population (P_o) of size N is created. Later this one is ordained, using the procedure of nondominated arranging. Then the usual binary tournament selection, recombination and mutation operators are used to create a new population (Q_o), of size N.

Main Loop

The NSGA-II procedure can be explained, by describing the th generation just as it is showed in Figure 1. The procedure begins with the combination of Pt and Qt forming a new population called Rt, then the population Rt is sorted using the nondomination criterion. Since all previous and current population members are included in Rt, elitism is ensure. The population Rt has a size of $2N$, later, the different fronts of nondominated solutions are created, being $F1$ the front that contains the better rank solutions. Figure 4 shows that, during the process of forming the new population $Pt+1$, the algorithm takes all members of the fronts $F1$ and $F2$, and some elements of the front $F3$; this is, because N solutions are needed exactly for the new population $Pt+1$ to find them exactly N solutions, the last front is ordained, which for this description is the number 3, arranging the solutions in descending order by means of the crowded comparison

($<n$), and selecting the best solutions needed to fill all population slots. After having the population $Pt+1$, the genetic operators of selection, crossing and mutation, are used to create the new population $Qt+1$ of size N. Finally it is mentioned that the selection process, the crowded comparison operator is used.

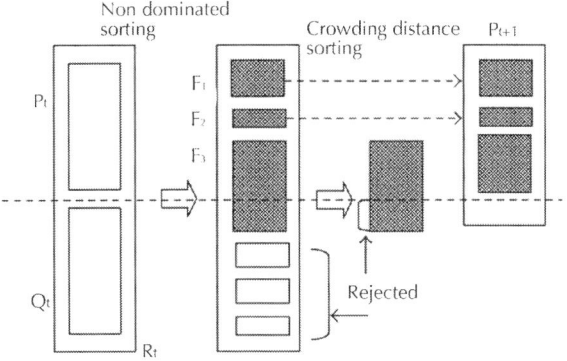

Figure 1: The NSGA II Procedure.

DESCRIPTION OF THE PROCESS

To demonstrate the effectiveness of NSGA-II on batch plant processes two examples are given here. The first example is about a batch plant consisting of three processing stages (mixer, reactor, and centrifuge) to manufacture two products, A and B. The second example treats four processing stages (mixer, reactor, extraction and centrifuge) to manufacture three products A, B and C. The data for examples 1 and 2 are illustrated, respectively, in Tables 1 and 2 (the processing times, size factor for the units and cost for each product).

Table 1: Data used in example 1

Processing times, $\tau_{ij}(h)$				Unit price for the product ($/Kg)		
Product	Mixer	Reactor	Centrifuge	Product	Cp	$C0$
A	8	20	4	A	0.35	0.08
B	10	12	3	B	0.37	0.1

Product	Size factors (L/kg)					
A	2	3	4			
B	4	6	3			
Cost of equipment ($, V in litres)				Minimum size = 250 L		
	$250 V^{0.6}$	$500 V^{0.6}$	$340 V^{0.6}$	Maximum size = 2500 L		
Operating cost factor (CE)						
	20	30	15			

Table 2: Data used in example 2

Processing times, τ_{ij} (h)					Unit price for the product ($/Kg)		
Product	Mixer	Reactor	Extractor	Centrifuge	Product	Cp	$C0$
A	1.15	9.86	0.4	0.5	A	0.27	0.08
B	5.95	7.01	0.7	0.42	B	0.29	0.10
C	3.96	6.01	0.85	0.3	C	0.32	0.12
γij	0.4	0.33	0.3	0.2			
Product	Size factors (L/kg)						
A	8.28	9.7	6.57	2.95			
B	5.58	8.09	6.17	3.27			
C	2.34	10.3	5.98	5.7			
Product	Coefficients Cij						
A	0.2	0.24	0.4	0.5			
B	0.15	0.35	0.7	0.42			

C	0.34	0.5	0.85	0.3			
Cost of equipment ($, V in litres)					Minimum size = 250 L		
	250 $V^{0.6}$	250 $V^{0.6}$	250 $V^{0.6}$	250 $V^{0.6}$	Maximum size = 10000 L		
Operating cost factor (CE)							
	20	30	15	30			

RESULTS AND DISCUSSION

Example 1

A four-objective optimization problem is considered and expressed in (1)–(4). The set of decision variables consists of the batch size, the total production time, the number of parallel units at each stage, the cycle time for each product, and the required volume of a unit in each stage. Since the number of parallel units at each stage is an integer decision variable, we code this variable as a binary variable. All other decision variables are coded as real numbers. Thus, there are 3 integer variables and 10 real variables. In addition to the constraints expressed by (5)–(8), we consider bounds on objective functions as additional constraints to generate feasible nondominated solutions in the range desired by the decision-maker, to have 19 constraints in all

$$f_i^L \leq f_i \leq f_i^U, \quad i = 1\ldots,4. \tag{10}$$

Then NSGA-II is employed to solve the optimization problem with the following parameters: maximum number of generation up to 200, population size 500, probability of crossover 0.85, probability of mutation 0.05, distribution index for the simulated crossover operation 10 and distribution index for the simulated mutation operation 20.

The Pareto-optimal solutions for example 1 are presented in Figure 2. The revenue (f_1) increases with the increase in operation cost (f_3), while the investment cost (f_2) decreases. When all the four objective functions are considered simultaneously, solutions obtained in the present study show improvement as by Huang and Wang [10] results for the same problem. For example, let us consider the solution presented by Huang and Wang [10] with unit reference membership level for all objectives: f_1=121350, f_2=171624, f_3=77299,4=5667. The solution (1) presented in Table 3 of the present study improves the above solution $f1,3,f_4$ while f_2 is comparable.

Table 3: Optimal objectif function values of example 1

Optimal objectif function values				
Case	f_1	f_2	f_3	f_4
1	122566.41	175955.39	73499.23	5579.37
2	124283.15	177389.07	74508.91	5582.87
3	126393.23	173354.67	84896.90	5638.60

Bounds for objective function:

$\left[f_1^L, f_1^U\right]=[110000,130000], \left[f_2^L, f_2^U\right]=[150000,20000], \left[f_3^L, f_3^U\right]=[60000,100000], \left[f_4^L, f_4^U\right]=[5500,6000]$

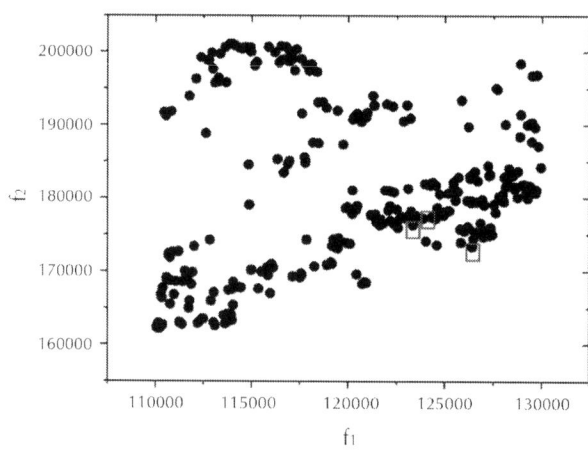

(a)

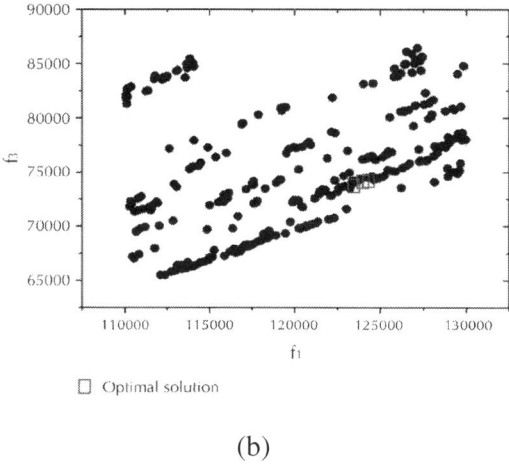

(b)

Figure 2: Pareto optimal solutions for example 1.

Figure 3 presents the relationships between some chosen decision variables. The large set of multiple optimal solutions provides the decision maker with immediate information about the relationship among the several objective criteria and a set of feasible solutions. Thus, it helps the decision-maker to select a highly confident choice of solution. The fixed optimal plant structure as 221 corresponds to a two mixers, two reactors, and one centrifuge design. The optimal solution is shown in Figure 4.

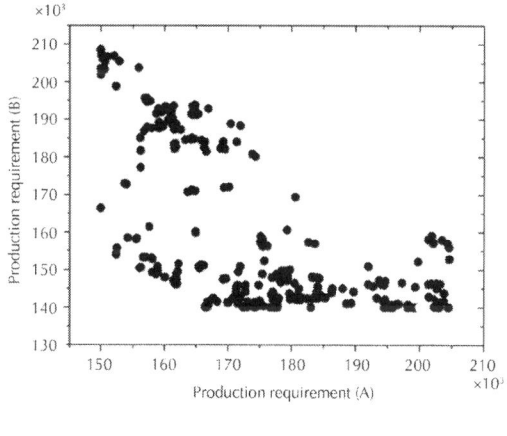

(a)

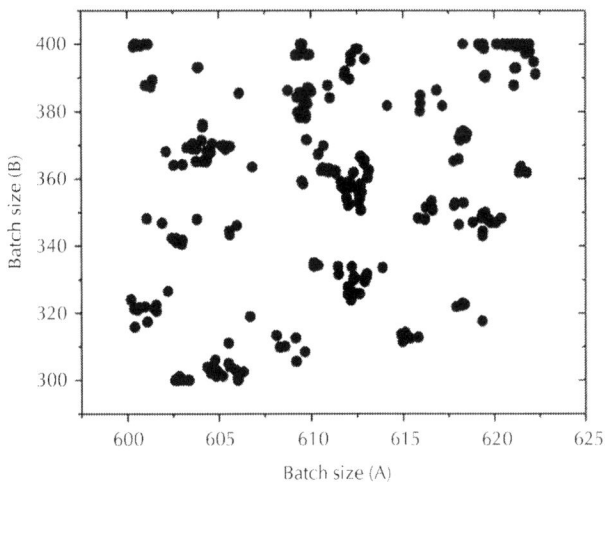

(b)

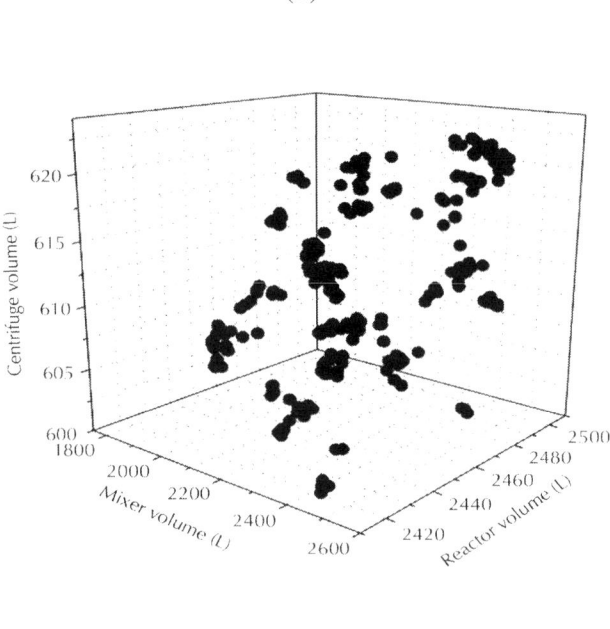

(c)

Figure 3: Relationships between some decision variables.

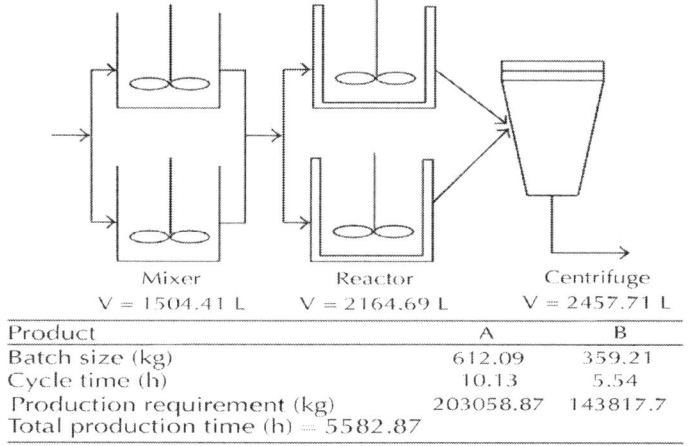

Product	A	B
Batch size (kg)	612.09	359.21
Cycle time (h)	10.13	5.54
Production requirement (kg)	203058.87	143817.7
Total production time (h) = 5582.87		

Figure 4: Optimal design of batch plant for example 1.

Example 2

The batch plant in this case consists of four processing stages to manufacture three products A, B, and C, with four-objective optimization problem as expressed in (1)–(4). The set of decision variables remains the same as that in example 1. But we deal with 4 integer variables, 14 real variables, and 31 constraints which includes bounds on objective functions.

The same model equations of example 1 are used here except the processing time, τ_{ij} in (7). The time required to process one batch of product i in stage j is expressed as:

$$\tau_{ij} = \overline{\tau}_{ij} + c_{ij} B_i^{y_j}, \quad \forall i, \forall j, \qquad (11)$$

where $\overline{\tau}_{ij} \geq 0$, $c_{ij} \geq 0$ and y_j are constants and B_i is the batch size for product i.

Thus, the processing time is not a constant, but depends on the decision parameters of the batch size. Table 2 presents the necessary data for the problem.

The constrained multiobjective MINLP problem is solved by NSGA-II with the same set of NSGA-II parameters as used in example 1.

As in example 1, the revenue (f_1) increases as operation cost (f_3) increases, while the investment cost (f_2) decreases following operation cost (f_3). The Pareto-optimal solutions for example 2 are presented in Figure 5. The relationships of the various decision variables are shown in Figure 6. Let mention that when all four objective functions are considered simultaneously, the solutions obtained in the present study improve significantly the results presented by Huang and Wang [10] for the same problem. For example, the solution presented by Huang and Wang [10] with unit reference membership level for all objectives (f_1 =274312, f_2=375688, f_3=175688, $f4$=5639) the solution (1) presented in Table 4 of the present study improves the above solution f_1, f_3, f_4 while f_2 is comparable.

Table 4: Optimal objectif function values of example 2

Optimal objectif function values				
Case	f_1	f_2	f_3	f_4
1	275766.10	388262.10	156449.20	5505.40
2	276096.30	369977.30	161392.30	5710.90
3	281818.70	369843.48	163552.80	5718.00

Bounds for objective function:

$[f_1^L, f_1^U] = [250000, 30000], [f_2^L, f_2^U] = [350000, 40000], [f_3^L, f_3^U] = [150000, 200000], [f_4^L, f_4^U] = [5500, 6000]$

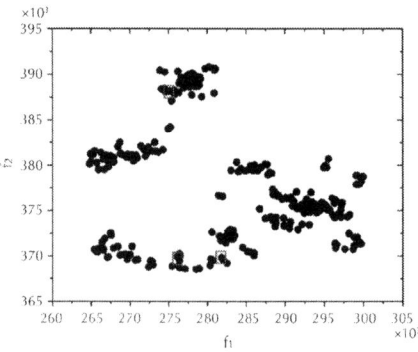

(a)

Optimal Solutions of Multiproduct Batch Chemical Process Using... 125

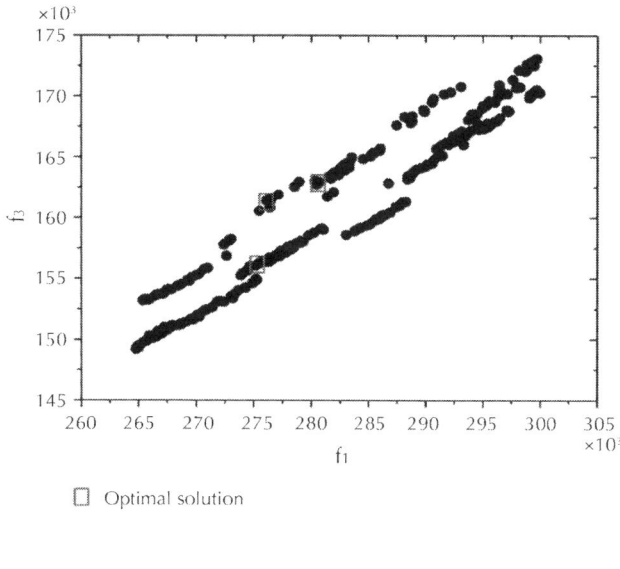

(b)

Figure 5: Pareto optimal solutions for example 2.

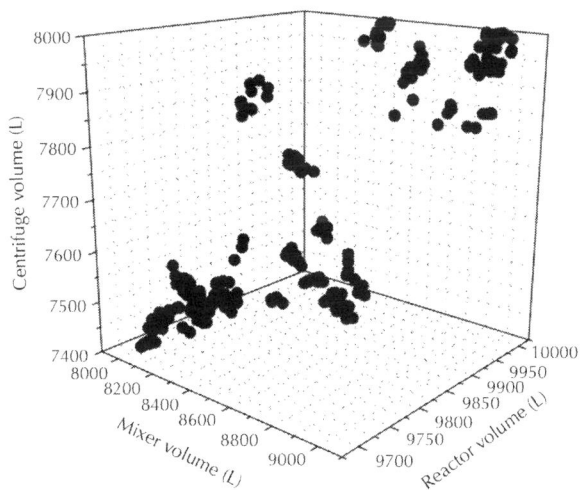

(a)

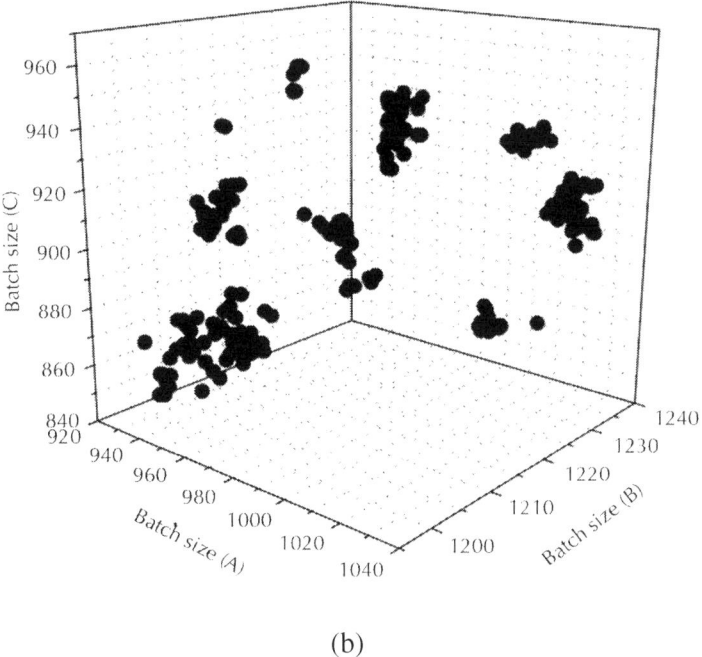

(b)

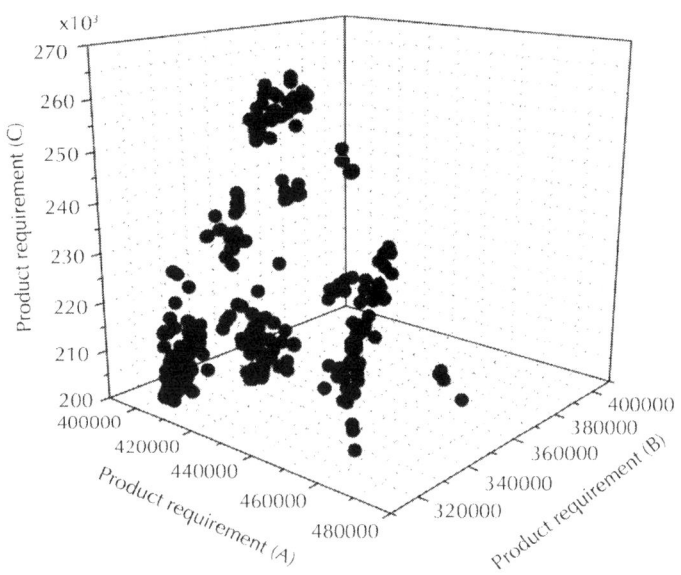

(c)

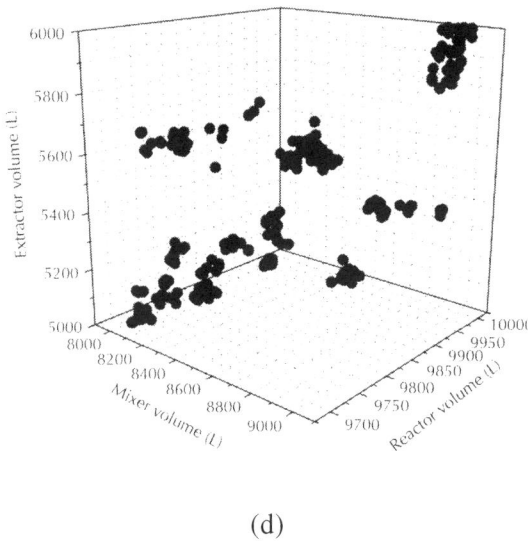

(d)

Figure 6: 3 Dimension plot of the relationship between some decision variables.

this example, the plant structure evolved as optimal is: two mixers, two reactors, two extractors, and one centrifuge as presented in Figure 7.

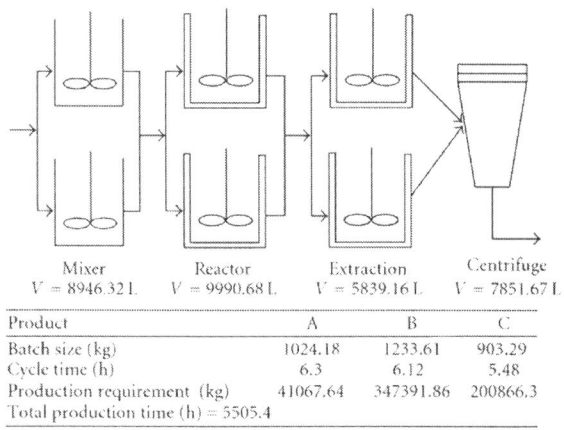

Product	A	B	C
Batch size (kg)	1024.18	1233.61	903.29
Cycle time (h)	6.3	6.12	5.48
Production requirement (kg)	41067.64	347391.86	200866.3
Total production time (h) = 5505.4			

Figure 7: Batch plant optimal design for example 2.

Implementation of a trade-off analysis is dependent upon the availability of the decision-maker's preferences.

CONCLUSIONS

A multiobjective decision in a batch plant process design is considered and a non dominating sorting genetic algorithm (NSGA-II) is developed to get an optimal zone containing solutions under the concept of Pareto set. NSGA-II capability has been proved in evolving the entire set of nondominating solutions along the Pareto front in a single run of the algorithm. Thus, the Decision Maker (DM) is provided with the best trade-off operating zone. Furthermore, a better confident choice of design among several compromises of the decision maker can be achieved if the decision variables effects on the objective functions are known.

Finally, the large set of solutions presents a useful base for further alternative approaches to fulfil the DM targets.

The inherent dynamic nature of batch processes allows for their ability to handle variations in feedstock and product specifications and provides the flexibility required for multiproduct or multipurpose facilities. They are thus best suited for the manufacture of low-volume, high-value products, such as specialty chemicals, pharmaceuticals, agricultural, food, and consumer products, and most recently the constantly growing spectrum of biotechnology-enabled products. Reduced time to market, lower production costs, and improved flexibility are all critical success factors for batch processes.

REFERENCES

1. A. Lamghabbar, S. Yacout, and M. S. Ouali, "Concurrent optimization of the design and manufacturing stages of product development," *International Journal of Production Research*, vol. 42, no. 21, pp. 4495–4512, 2004.
2. D.-M. Cao and X.-G. Yuan, "Optimal design of batch plants with uncertain demands considering switch over of operating modes of parallel units," *Industrial and Engineering Chemistry Research*, vol. 41, no. 18, pp. 4616–4625, 2002.

3. W. Chunfeng and Z. Xin, "Ants foraging mechanism in the design ofmultiproduct batch chemical process," *Industrial and Engineering Chemistry Research*, vol. 41, no. 26, pp. 6678–6686,2002.
4. H. D. Goel, M. P. C. Weijnen, and J. Grievink, "Optimal reliable retrofit design of multiproduct batch plants," *Industrial and Engineering Chemistry Research*, vol. 43, no. 14, pp. 3799–3811, 2004.
5. S.-K. Heo, K.-H. Lee, H.-K. Lee, I.-B. Lee, and J. H. Park, "A new algorithm for cyclic scheduling and design of multipurpose batch plants," *Industrial and Engineering Chemistry Research*, vol. 42, no. 4, pp. 836–846, 2003.
6. J. M. Montagna, A. R. Vecchietti, O. A. Iribarren, J. M. Pinto, and J. A. Asenjo, "Optimal design of protein production plants with time and size factor process models," *Biotechnology Progress*, vol. 16, no. 2, pp. 228–237, 2000.
7. A. Chakraborty, A. Malcolm, R. D. Colberg, and A. A. Linninger, "Optimal waste reduction and investment planning under uncertainty," *Computers & Chemical Engineering*, vol. 28, no. 6-7, pp. 1145–1156, 2004.
8. L. Cavin, U. Fischer, F. Glover, and K. Hungerb¨uhler, "Multiobjective process design in multi-purpose batch plants using a Tabu Search optimization algorithm," *Computers & Chemical Engineering*, vol. 28, no. 4, pp. 459–478, 2004.
9. T. Pinto, A. P. F. D. Barbosa-P´ovoa, and A. Q. Novais, "Optimal design and retrofit of batch plants with a periodic mode of operation," *Computers & Chemical Engineering*, vol. 29, no. 6, pp. 1293–1303, 2005.
10. H.-J. Huang and F.-S. Wang, "Fuzzy decision-making design of chemical plant using mixed-integer hybrid differential evolution," *Computers & Chemical Engineering*, vol. 26, no. 12, pp. 1649–1660, 2002.
11. S. Dedieu, L. Pibouleau, C. Azzaro-Pantel, and S. Domenech,"Design and retrofit of multiobjective batch plants via a multicriteria genetic algorithm," *Computers & Chemical Engineering*,vol. 27, no. 12, pp. 1723–1740, 2003.

12. M. Solimanpur, P. Vrat, and R. Shankar, "A multi-objective genetic algorithm approach to the design of cellular manufacturing systems," *International Journal of Production Research*, vol. 42, no. 7, pp. 1419–1441, 2004.
13. C. Dimopoulos, "Multi-objective optimization of manufacturing cell design," *International Journal of Production Research*, vol. 44, no. 22, pp. 4855–4875, 2006.
14. J.-P. Brans and B. Mareschal, "The PROMCALC and GAIA decision support system for multicriteria decision aid," *Decision Support Systems*, vol. 12, no. 4-5, pp. 297–310, 1994.
15. V. Bhaskar, S. K. Gupta, and A. K. Ray, "Applications of multiobjective optimization in chemical engineering," *Reviews in Chemical Engineering*, vol. 16, no. 1, pp. 1–54, 2000.
16. C. Fonseca and P. Fleming, "An overview of evolutionary algorithms in multiobjective optimization," *Evolutionary Computation*, vol. 3, no. 1, pp. 1–16, 1993.
17. J. Schaffer, "Multiple objective optimization with vector evaluated genetic algorithms," in *Proceedings of the 1st International Conference on Genetic Algorithms and Their Applications (ICGA '85)*, pp. 93–100, Pittsburgh, Pa, USA, July 1985.
18. H. Ishibuchi and T. Murata, "A multi-objective genetic local search algorithm and its application to flowshop scheduling," *IEEE Transactions on Systems, Man and Cybernetics, Part C*, vol. 28, no. 3, pp. 392–403, 1998.
19. D. E. Goldberg, *Genetic Algorithms in Search, Optimization and Machine Learning*, Addison-Wesley, Reading, Mass, USA, 1989.
20. J. Horn, N. Nafpliotis, and D. E. Goldberg, "A niched Pareto genetic algorithm for multiobjective optimization," in *Proceedings of the 1st IEEE Conference on Evolutionary Computation, IEEE World Congress on Computational Intelligence (ICEC '94)*, vol. 1, pp. 82–87, Orlando, Fla, USA, June 1994.
21. E. Zitzler and L. Thiele, "Multiobjective evolutionary algorithms: a comparative case study and the strength Paretom approach," *IEEE Transactions on Evolutionary Computation*, vol. 3, no. 4, pp. 257–271, 1999.
22. K. C. Tan, T. H. Lee, and E. F. Khor, "Evolutionary algorithms with dynamic population size and local exploration for multiobjective

optimization," *IEEE Transactions on Evolutionary Computation*, vol. 5, no. 6, pp. 565–588, 2001.

23. N. Srinivas and K. Deb, "Muiltiobjective optimization using nondominated sorting in genetic algorithms," *Evolutionary Computation*, vol. 2, no. 3, pp. 221–248, 1994.

24. K. Deb, A. Pratap, S. Agarwal, and T. Meyarivan, "A fast and elitist multiobjective genetic algorithm: NSGA-II," *IEEE Transactions on Evolutionary Computation*, vol. 6, no. 2, pp. 182–197, 2002.

Chapter 6

Bio and Chemical Sensors Based on Surface Plasmon Resonance in a Plastic Optical Fiber

Nunzio Cennamo[1] and Luigi Zeni[1]

[1]Department of Industrial and Information Engineering, Second University of Naples, Aversa, Italy

INTRODUCTION

Surface Plasmon Resonance (SPR) is known to be a very sensitive technique for determining refractive index variations at the interface between a metallic layer and a dielectric medium (analyte). SPR is widely used as a detection principle for many sensors operating in different application fields, such as bio and chemical sensing. In practical implementations, the biological targets are usually transported through a microfluidic system by means of a buffer fluid or a carrier fluid. In SPR sensors, the transducing media (ligands) are usually bonded on the metallic layer surface so that, when they react with the target molecules present in the analyte, the refractive index at the outer interface changes, and this change is detected by suitable optical interrogation. In the scientific literature, many different configurations based on SPR in silica optical fibers, are usually found [1,2].

In general, the optical fiber employed is either a glass one or a plastic one (POF). For low-cost sensing systems, POFs are especially advantageous due to their excellent flexibility, easy manipulation, great numerical aperture, large diameter, and the fact that plastic is able to withstand smaller bend radii than glass. The advantages of using POFs is that the properties of POFs, that have increased their popularity and competitiveness for telecommunications, are exactly those that are important for optical sensors based on optical fibers. Moreover, a further advantage of POF sensors is that they are simpler to manufacture than those made using silica optical fibers. In the scientific literature only simple POF sensors, based on laterally polished bent sections prepared along a plastic optical fiber, can be usually found.

In this chapter, POF sensor configurations are presented in order to monitor an aqueous environment (refractive index around 1.333), with a resolution ranging from 10^{-4} to 10^{-3} (RIU). The classic geometries of sensors based on SPR in silica optical fiber are adapted and borrowed for POF, so representing a simple approach to low cost plasmonic sensing.

The planar gold layer of the sensors, the low resolution and refractive index ranging from 1.332 to 1.420 are three good factors for forthcoming bio/chemical sensors implementation.

Another aspect to be considered is that most often the SPR bio-chemical sensor system is based on a high refractive index prism coated with a thin metallic layer. The incidence angle of the light can be changed in a wide range and, as a consequence, the surface plasma waves (plasmons) may exist whatever the surrounding medium, i.e. a gas or a liquid. These sensors are usually bulky and require expensive optical equipment, not easy to be miniaturized. In addition, the remote sensing can be very difficult to exploit. On the contrary, the use of a POF makes the remote sensing straightforward, and may reduce the cost and dimension of the device, with the possibility of integration of SPR sensing platform with optoelectronic devices, eventually leading to "Lab-on-a-chip".

SPR PHENOMENON

In the optical phenomenon of Surface Plasmon Resonance, a metal-dielectric interface supports a p-polarized electromagnetic wave, namely Surface Plasmon Wave (SPW), which propagates along the interface. When the p-polarized light is incident on this metal-dielectric interface in such a way that the propagation constant (and energy) of resultant evanescent wave is equal to that of the SPW, a strong absorption of light takes place as a result of transfer of energy and the output signal exhibits a sharp dip at a particular wavelength known as the resonance wavelength. The so-called resonance condition is given by the following expression:

$$K_0 n_c \sin\vartheta = K_0 \left(\frac{\varepsilon_{mr} n_s^2}{\varepsilon_{mr} + n_s^2}\right)^{1/2}; \quad K_0 = \frac{2\pi}{\lambda} \qquad (1)$$

The term on the left-hand side is the propagation constant K_{inc} of the evanescent wave generated as a result of Attenuated Total Reflection (ATR) of the light incident at an angle θ through a light coupling device (such as prism or optical fiber) of refractive index n_c. The right-hand term is the SPW propagation constant (K_{SP}) with ε_{mr} as the real part of the metal dielectric constant (ε_m) and n_s as the refractive index of the sensing (dielectric) layer. This matching condition of propagation constants is heavily sensitive to even a slight change in the dielectric constant, which makes this technique a powerful tool for sensing of different parameters.

Spectral Mode Operation

In SPR sensors with spectral interrogation, the resonance wavelength (λ_{res}) is determined as a function of the refractive index of the sensing layer (n_s). If the refractive index of the sensing layer is altered by δ_{ns}, the resonance wavelength shifts by $\delta\lambda_{res}$. The sensitivity (S_n) of an SPR sensor with spectral interrogation is defined as [3]:

$$S_n = \frac{\delta\lambda_{res}}{\delta n_s} \left[\frac{nm}{RIU}\right] \qquad (2)$$

Owing to the fact that the vast majority of the field of an SPW is concentrated in the dielectric, the propagation constant of the SPW is extremely sensitive to changes in the refractive index of the dielectric itself. This property of SPW is the underlying physical principle of affinity SPR bio/chemical sensors. In the case of artificial receptors, as molecular imprinted polymers (MIPs), the polymeric film on the surface of metal selectively recognizes and captures the analyte present in a liquid sample so producing a local increase in the refractive index at the metal surface. The refractive index increase gives rise to an increase in the propagation constant of SPW propagating along the metal surface which can be accurately measured by optical means. The magnitude of the change in the propagation constant of an SPW depends on the refractive index change and its overlap with the SPW field. If the binding occurs within the whole depth of the SPW field, the binding-induced refractive index change produces a change in the real part of the propagation constant, which is directly proportional to the refractive index change.

The resolution (Δn) of the SPR-based optical sensor can be defined as the minimum amount of change in refractive index detectable by the sensor. This parameter (with spectral interrogation) definitely depends on the spectral resolution ($\delta\lambda_{DR}$) of the spectrometer used to measure the resonance wavelength in a sensor scheme. Therefore, if there is a shift of $\delta\lambda\rho_{es}$ in resonance wavelength corresponding to a refractive index change of δn_s, then resolution can be defined as:

$$S_n = \frac{\delta \lambda_{res}}{\delta n_s} \left[\frac{nm}{RIU} \right] \quad (3)$$

The Signal-to-Noise Ratio of an SPR sensor depends on how accurately and precisely the sensor can detect the resonance wavelength and hence, the refractive index of the sensing layer. This accuracy in detecting the resonance wavelength further depends on the width of the SPR curve.

The narrower the SPR curve, the higher the detection accuracy. Therefore, if $\delta \lambda_{SW}$ is the spectral width of the SPR response curve corresponding to some reference level of transmitted power, the detection accuracy of the sensor can be assumed to be inversely proportional to $\delta \lambda_{SW}$.

The signal-to-noise ratio of the SPR sensor with spectral interrogation is, thus, defined as:

$$SNR(n) = \left(\frac{\delta \lambda_{res}}{\delta \lambda_{SW}} \right)_n \quad (4)$$

where $\delta \lambda_{SW}$ can be calculated as the full width at half maximum of the SPR curve (FWHM). SNR is a dimensionless parameter strongly dependent on the refractive index changes.

Amplitude Mode Operation

In SPR sensors, it can be used a simpler scheme with a monochromatic light source and an optical power meter in the amplitude mode operation, because in the shoulders of spectral response curve, appreciable intensity differences are present, as the given refractive index changes, due to the shift of the curve itself.

In amplitude mode operation the sensitivity (S_n) of an SPR sensor is defined as:

$$S_n = \frac{\delta I_{norm}}{\delta n_s} \left[\frac{a.u.}{RIU}\right] \qquad (5)$$

where *Inorm* is the relative output, normalized to a reference level, in order to compensate for light source fluctuations.

The resolution (Δn) of the SPR-based optical sensor, in amplitude mode operation, can be defined as:

$$\Delta n = \frac{\delta n_s}{\delta I_{norm}} \sigma [RIU] \qquad (6)$$

where σ is the standard deviation of the relative output.

A POF SENSOR SYSTEM FOR BIOSENSOR IMPLEMENTATION

The fabricated optical sensor system was realized removing the cladding of a plastic optical fiber along half the circumference, spin coating on the exposed core a buffer of Microposit S1813 photoresist, and finally sputtering a thin gold film using a sputtering machine [4].

The plastic optical fiber has a PMMA core of 980 μm and a fluorinated polymer cladding of 20 μm. The experimental results indicate that the configuration with a fiber diameter of 1000 μm exhibits better performance in terms of sensitivity and resolution but not in terms of SNR [5], as shown in section 5.2.

The refractive index, in the visible range of interest, is about 1.49 for PMMA, 1.41 for fluorinated polymer and 1.61 for Microposit S1813 photoresist. The sample consisted in a plastic optical fiber without jacket embedded in a resin block, with the purpose of easing the polishing process. The polishing process was carried out with a 5 μm polishing paper in order to remove the cladding and part of the core. After 20 complete strokes following a "8-shaped" pattern (according to the manufacturer recommendations, as shown in Figure 1) in order to completely expose the core, a 1 μm polishing paper was used for another 20 complete strokes following a "8-shaped" pattern. The realized sensing region was about 10 mm in length.

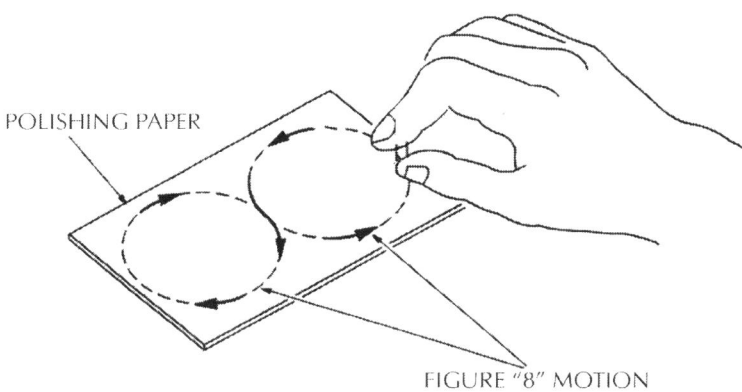

Figure 1: "8-shaped" pattern for POF polishing.

The buffer of Microposit S1813 photoresist was realized by means of spin coating. The Microposit S1813 photoresist is deposited in one drop (about 0.1 ml) on the center of the substrate. The sample is then spun at 6,000 rpm for 60 seconds. The final thickness of photoresist buffer was about 1.5 μm.

As it will be shown in the section 5.1, the experimental results indicate that this configuration with the photoresist buffer layer exhibits better performance in terms of detectable refractive index range and SNR [4].

Finally, a thin gold film was sputtered by using a sputtering machine (Bal-Tec SCD 500). The sputtering process was repeated twice with a current of 60 mA for a time of 35 seconds (20 nm for step). The gold film so obtained was 60 nm thick and presented a good adhesion to the substrate, verified by its resistance to rinsing in de-ionized water.

This sensor based on SPR in a POF is a common tool for surface interaction analysis and biosensing, widely used as a detection principle for sensors that operate in different areas of bio and chemical sensing as reported in several recent review papers [6,7]. In this case on the gold surface there is a bio or chemical layer for the selective detection and analysis of analyte in aqueous solution (see figure 2).

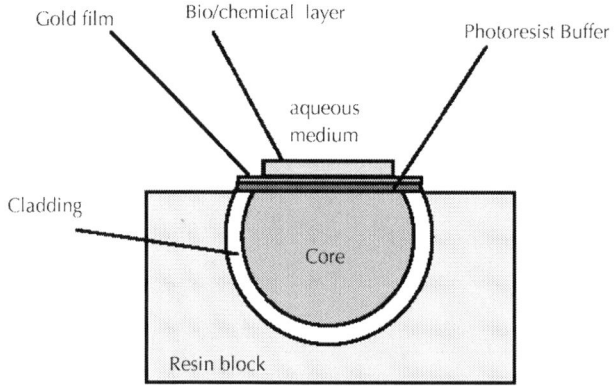

Figure 2: Section of POF sensor.

EXPERIMENTAL CONFIGURATIONS FOR SPR SENSORS IN PLASTIC OPTICAL FIBERS

The experimental measurements for the characterization of the POF sensor, presented in the previous section, were carried out in two different ways: spectral and amplitude mode. Figure 3 shows the experimental setup arranged to measure the transmitted light spectrum and was characterized by a halogen lamp, illuminating the optical sensor system, and a spectrum analyzer. The employed halogen lamp exhibits a wavelength emission range from 360 nm to 1700 nm, while the spectrum analyzer detection range was from 200 nm to 850 nm. The spectral resolution of the spectrometer was 1.5 nm (FWHM).

Bio and Chemical Sensors Based on Surface Plasmon Resonance... 141

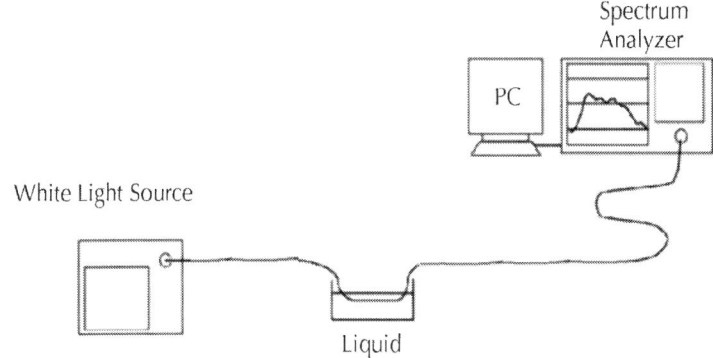

Figure 3: Setup to measure the transmitted light spectrum.

Figure 4 shows the measurements carried out, obtaining SPR transmission spectra, normalized to the spectrum achieved with air as the surrounding medium, for three different water-glycerin solutions with refractive index 1.333, 1.351, 1.371, respectively. The sensitivity is calculated as the slope of the resonance wavelength versus refractive index curve, for three refractive index values. In figure 5 the experimental data and the linear fitting are presented.

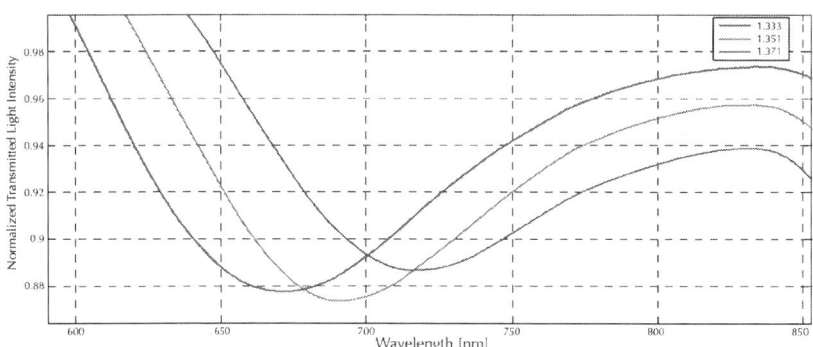

Figure 4: Normalized transmitted light intensity as function a function of the wavelength.

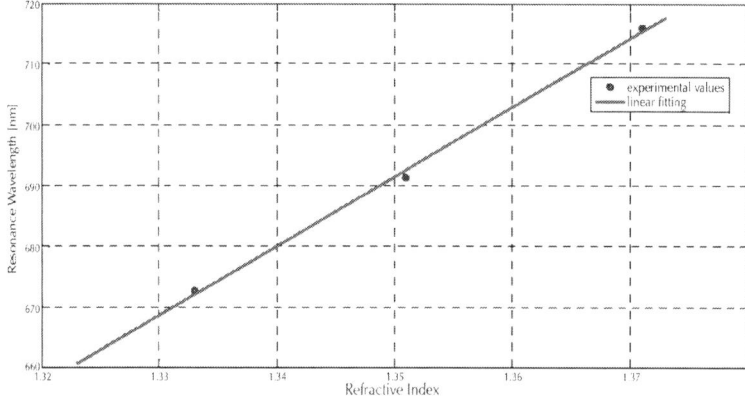

Figure 5: Resonance wavelength as a of the refractive index.

Figure 6 shows the second setup. It is composed of an LED, whose wavelength is 670nm, as the light source, a beam splitter, two photodiodes whose function is to convert the light into an electrical signal and an oscilloscope for signal acquisition connected to a PC.

In figure 7, the relative output is plotted as a function of time for three different refractive index values.

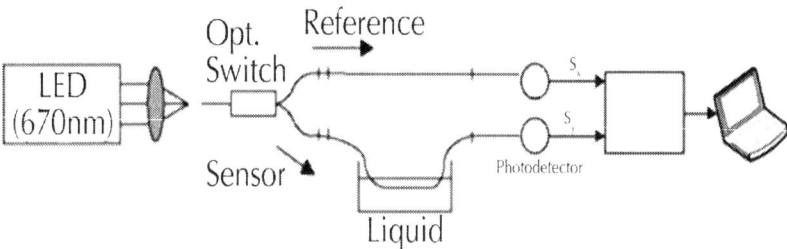

Figure 6: Setup using a LED and two photodiode.

According to an experiment for the evaluation of system stability, the standard deviation of the relative output was found to be 3.84×10^{-4} for the 8 min continuous operation in the air with the intention of circumventing liquid evaporation. In figure 8, the relative output for three different refractive index and the linear fitting to the experimental values are presented.

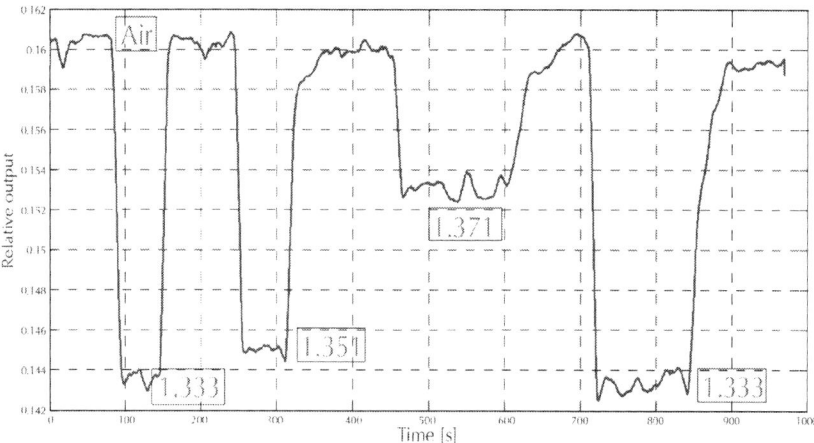

Figure 7: Relative output as a function of time, for different refractive index values.

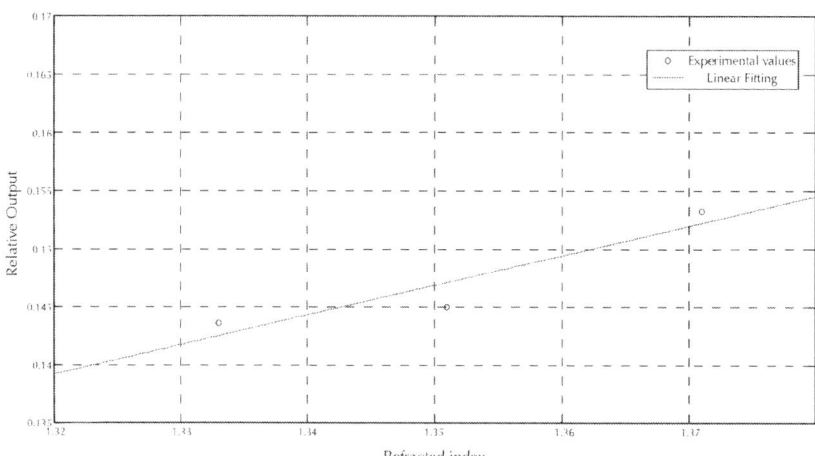

Figure 8: Relative output as a function of refractive index.

For a clearer comparative analysis of the two experimental setups, table 1 summarizes the average values of the experimentally measured performance parameters, evaluated by Matlab software, for external medium refractive index ranging from 1.333 to 1.371.

Table 1. Performance parameters for different experimental configurations

Experimental setup	Sensitivity (S)	Margin of error (∂E)	$\Delta n = S{-}1 * \partial E$
White Light Source/ spectrum analyzer	δ resonance wavelengthδ refractive index=1.14* 10^3 [nmRIU]	(spectral resolution of the spectrometer) 1.5 nm (FWHM)	0.00131 [RIU]
LED/photodiodes	δ relative outputδ refractive index=0.26 [a.u.RIU]	(standard deviation) 3.84×10−4	0.00147 [RIU]

PERFORMANCE COMPARISON OF SENSORS BASED ON SPR IN POF

In this section different POF sensor configurations are presented and experimentally tested with spectral interrogation: First, the configurations with and without photoresist buffer layer; then, the configurations with two different POF core diameters and finally the configuration with a tapered POF.

POF Sensors With and Without the Photoresist Layer

In this section we present two configurations with and without the photoresist buffer layer (see figure 2). In the series of performed experiments, water-glycerin solutions were used to achieve an aqueous medium with variable refractive index. Without the buffer layer, in the same operating conditions, the sensor is capable of monitoring refractive indexes ranging from 1.330 to 1.360. In Figure 9 are presented the experimentally obtained SPR transmission spectra, normalized to the spectrum achieved with air as the surrounding medium, for three different water-glycerin solutions with refractive index ranging from 1.332 to 1.352.

In the presence of the photoresist buffer layer, the refractive index range is increased. In particular, this fiber optic sensor is capable of monitoring an aqueous environment whose refractive index ranges from 1.332 to 1.418.

In Figure 10 are presented the experimentally obtained SPR transmission spectra, normalized to the spectrum achieved with air as the surrounding medium, obtained in this case with the photoresist buffer layer, for different water-glycerin solutions with refractive index ranging from 1.332 to 1.418.

In presence of the photoresist buffer layer, the refractive index range is increased while the sensitivity is the same.

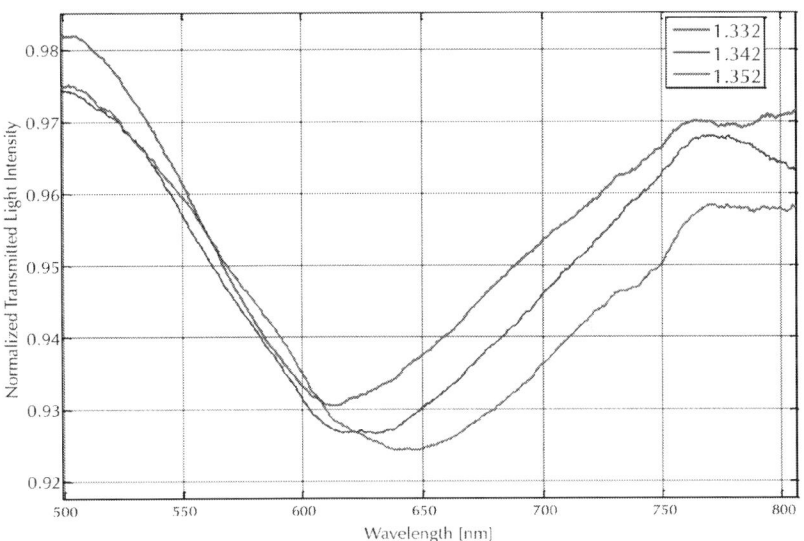

Figure 9: Experimentally obtained SPR transmission spectra, normalized to the air spectrum, for different refractive index of the aqueous medium. Configuration without the photoresist buffer layer.

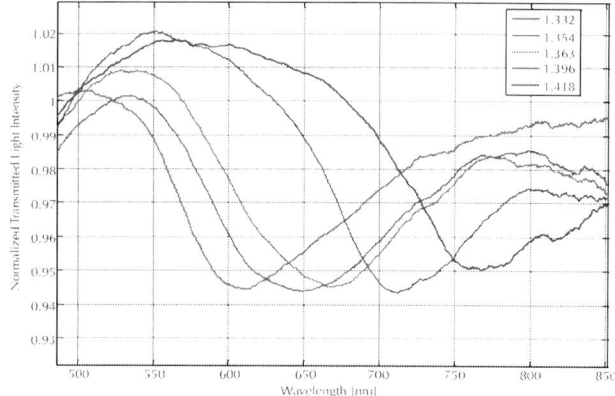

Figure 10: Experimentally obtained SPR transmission spectra, normalized to the air spectrum, for different refractive index of the aqueous medium. Configuration with the photoresist buffer layer.

Without the photoresist buffer layer there is a decrease in the power transmitted to the fiber end facet, due to a greater dissipation. This decrease results in the decrease of the SPR curve and the increase of the SPR curve width. Therefore, it can be conveniently established that SPR curve width increases ($\delta\lambda_{SW}$) without the photoresist buffer layer, as shown for example in Figure 11 for a refractive index equal to 1.332 (water).

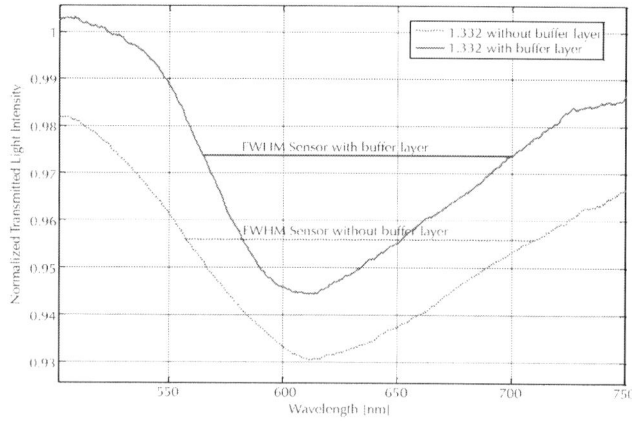

Figure 11: The full width at half maximum (FWHM) of the SPR curve for the two sensors configurations with and without the buffer layer for an external refractive index of 1.332.

The observed absorption band is the result of the convolution of different resonance peaks. Each peak is obtained for a specific resonance condition defined by a given angle-wavelength couple. Therefore, the experimental results indicate that the configuration with the photoresist buffer layer exhibits better performance in terms of detectable refractive index range and SNR.

POF Sensors with Two Different POF Core Diameters

In this section we presented the influence of POF core diameter on sensor performances. Before entering the details of the discussion, as a similar analysis is present in the literature with reference to a sensor based SPR in silica optical fiber without any buffer layer between the fiber core and the gold film [8], it is convenient to briefly recall some fundamental aspects of light rays propagation in optical fibers where surface plasmons are excited.

Inside an optical fiber, any light ray making an angle from the normal to core-cladding interface undergoes multiple reflections (N_{ref}), depending on the length of SPR sensing region (L) and fiber core diameter (D), according to the following relation [8]:

$$N_{ref}(\theta) = \frac{L}{D \tan \theta} \quad (7)$$

To determine the effective transmitted power, the reflectance (R_e) for a single reflection is raised to the power equal to corresponding number of reflections. Therefore, the generalized expression (all guided rays) for the normalized transmitted power (P_{trans}) in sensors based on SPR in fiber optic can be written as:

$$P_{trans} = \frac{\int_{\theta cr}^{\pi/2} R_e^{N_{ref}(\theta)} I(\theta) d\theta}{\int_{\theta cr}^{\pi/2} I(\theta) d\theta} \tag{8}$$

In Equation (8), I(θ) is the angular intensity distribution corresponding to the light source used. Further, θχρ is the critical angle of the fiber, which heavily depends on the Numerical Aperture (NA) of the fiber and light wavelength.

The angular range from θ_{cr} to ϖ/2 covers whole range of guided rays (or modes) as these angles correspond to the highest order mode and the fundamental mode of an optical fiber, respectively. The number of modes that can propagate in a fiber depends on the fiber's Numerical Aperture (or acceptance angle) as well as on its core diameter and the wavelength of the light. For a step-index multimode fiber, the number of such modes, M, is approximated (M >> 1) by:

$$M \cong 0.5 * \left(\frac{\pi * D * NA}{\lambda}\right)^2 \tag{9}$$

where D is the core diameter, λ is the operating wavelength, NA is the Numerical Aperture (or acceptance angle, as shown in Figure 12).

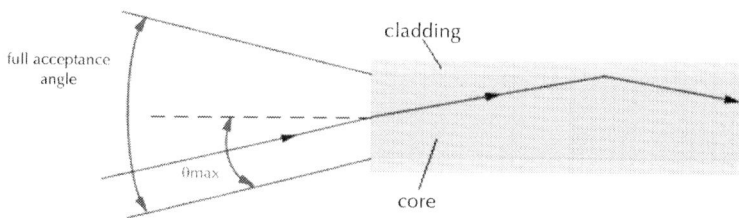

Figure 12: Optical fiber acceptance cone.

In general, numerical aperture of a Plastic Optical Fiber is greater than that of a Silica Optical Fiber. The resonance condition (see Equation (1)) is satisfied at different wavelengths depending on which combination of core diameter and sensing region length is considered. It is so clear that the performance parameters of a fiber optic SPR sensor strictly depend on the values of design parameters such as fiber core diameter (D), sensing region length (L), and numerical aperture (NA). In this section, we analyze the influence of two values of Plastic Optical Fiber core diameter (D) on the performance of a sensor based on Surface Plasmon Resonance in a POF, where the sensing region length is fixed and a photoresist buffer layer is placed between the fiber core and the gold film (seefigure 2).

For sensors based on SPR in optical fiber (silica or plastic) the shift in resonance wavelength ($\delta\lambda_{res}$), for a fixed refractive index variation (δn_s), increases with a decrease in the number of reflections. Therefore, sensitivity increases with the increase of fiber core diameter and with the decrease of sensing region length.

Figure 13 reports the experimentally obtained SPR transmission spectra, when the diameter POF is 1,000 μm (figure 13 (a)) and 250 μm (figure 13 (b)), normalized to the spectrum achieved with air as the surrounding medium, for five different water-glycerin solutions with refractive index ranging from 1.332 to 1.372.

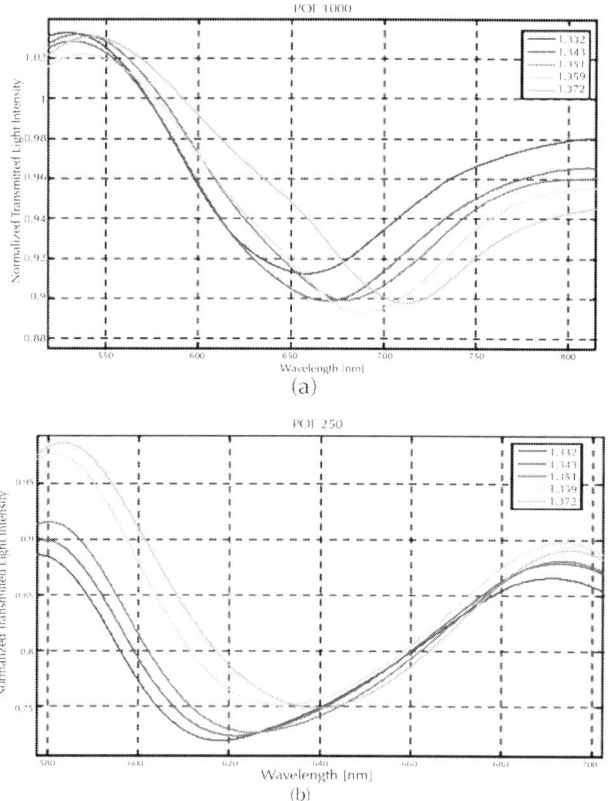

Figure 13: Experimentally obtained SPR transmission spectra, normalized to the air spectrum, for different refractive index of the aqueous medium. (a) Configuration with a 1,000 μm diameter POF; (b) Configuration with a 250 μm diameter POF

Figure 14 shows the resonance wavelength versus the refractive index obtained with the two different configurations. In the same figure is also presented the linear fitting to the experimental data, showing a good linearity for the sensors. The Pearson's correlation coefficient (R) is 0.99 for the sensor with a POF of 1,000 μm and 0.98 for the sensor with a POF of 250 μm diameter.

The sensitivity, as defined in Equation (2), is the shift of the resonance wavelength (nm) per unit change in refractive index (nm/RIU). Therefore, it is the angular coefficient of the linear fitting. Figure 14 shows as the sensitivity increases with the fiber core diameter.

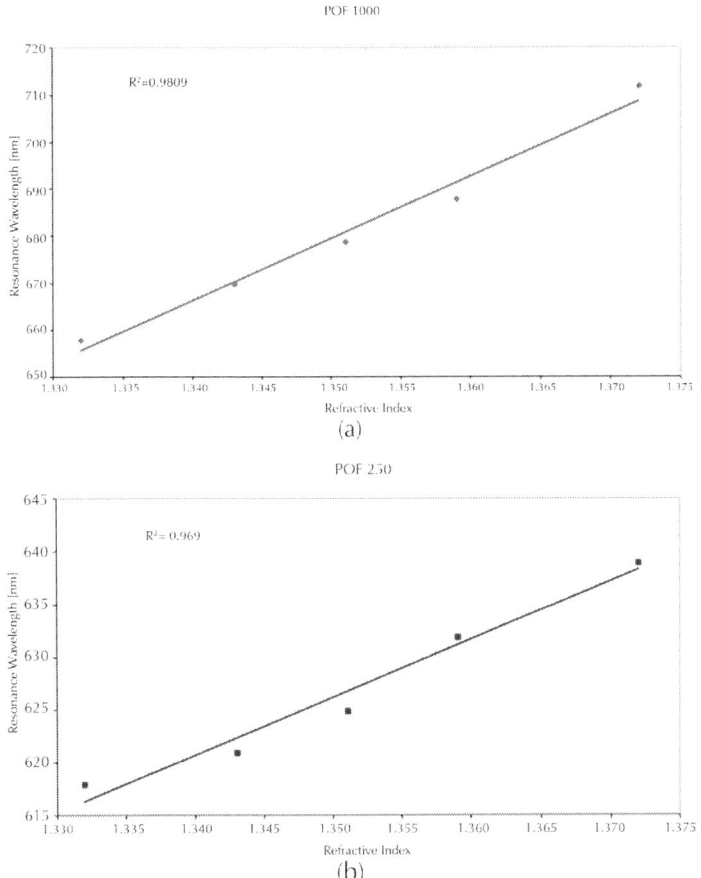

Figure 14: Plasmon resonance wavelength as a function of the refractive index. (a) Configuration with a 1,000 μm diameter POF. (b) Configuration with a 250 μm diameter POF.

Furthermore, as sensor's resolution also depends on the variation of $\delta\lambda\rho_{es}$ (see Equation (3)), therefore, similarly to sensitivity, resolution also tends to improve for larger fiber core diameters (see Figure 15). In fact, the resolution (Δn) can be calculated as the angular coefficient of the linear fitting in Figure 15 multiplied to the spectral resolution ($\delta\lambda_{DR}$) of the spectrometer used to measure the resonance wavelength.

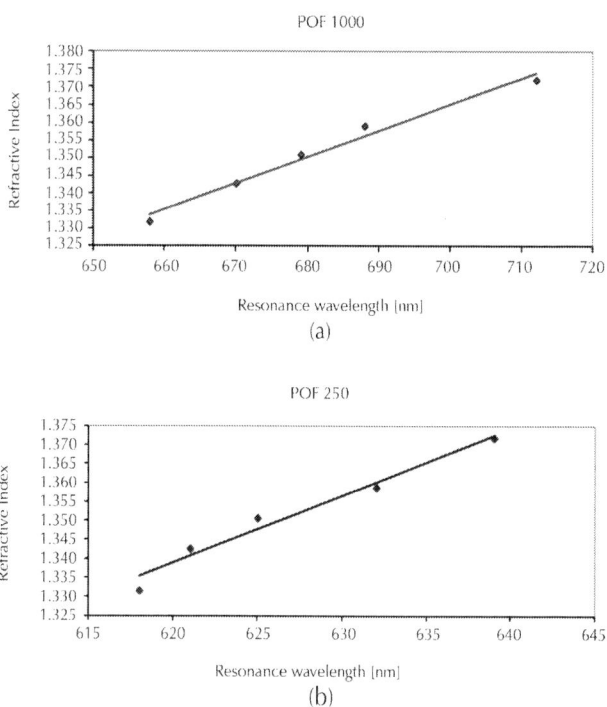

Figure 15: Refractive index as a function of the plasmon resonance wavelength. (a) Configurations with a 1,000 μm diameter POF. (b) Configurations with a 250 μm diameter POF.

The experimental results obtained with the two values of POF core diameter have shown as the Numerical Aperture of POF and the photoresist buffer layer have produced a different behavior with respect to many different configurations based on SPR in silica optical fibers, as SNR is concerned. From our experimental results [5], it is clear that the shift in resonance wavelength ($\delta\lambda_{res}$), for a fixed refractive index variation (δ_{ns}), increases when the core diameter increases. Therefore, sensitivity increases with an increase in fiber core diameter. Furthermore, in the sensors based on SPR in POF (configuration with the photoresist buffer layer) as already established, SPR curve width ($\delta\lambda\Sigma_W$) increases with an increase in fiber core diameter. Therefore, it can be conveniently established that SPR curve width increases ($\delta\lambda\Sigma_W$) with the increase of fiber core diameter, as shown in Figure 16 for a refractive index equal to 1.332.

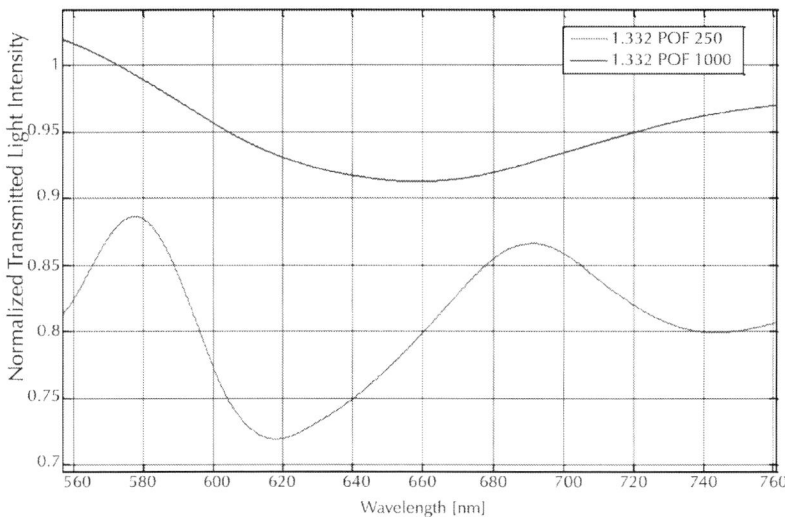

Figure 16: The full width at half maximum of the SPR curve for the two sensors configurations (250 μm and 1,000 μm POF diameter) for an external refractive index of 1.332.

SPR curve width $\delta\lambda_{SW}$ can be calculated as the full width at half maximum (FWHM) of the SPR curve. FWHM of the SPR curve as a function of the refractive index is shown in Figure 17. Therefore, SNR is expected to decrease because an increase in the shift in resonance wavelength ($\delta\lambda\rho_{es}$) produces a larger increase in the curve width ($\delta\lambda_{SW}$), for a fixed increase in fiber core diameter.

More precisely, for a POF with 250 μm of diameter, the angular coefficient of the linear fitting shown in Figure 14 ($\delta\lambda\rho_{es}$) is greater than the angular coefficient of the linear fitting shown in Figure 17($\delta\lambda_{SW}$). In this case SNR is greater than one. For a POF with 1,000 μm of diameter the angular coefficient of the linear fitting shown in Figure 14 ($\delta\lambda\rho_{es}$) is lower than the angular coefficient of the linear fitting shown in Figure 17 ($\delta\lambda\Sigma_{W}$). In this case SNR is less than one.

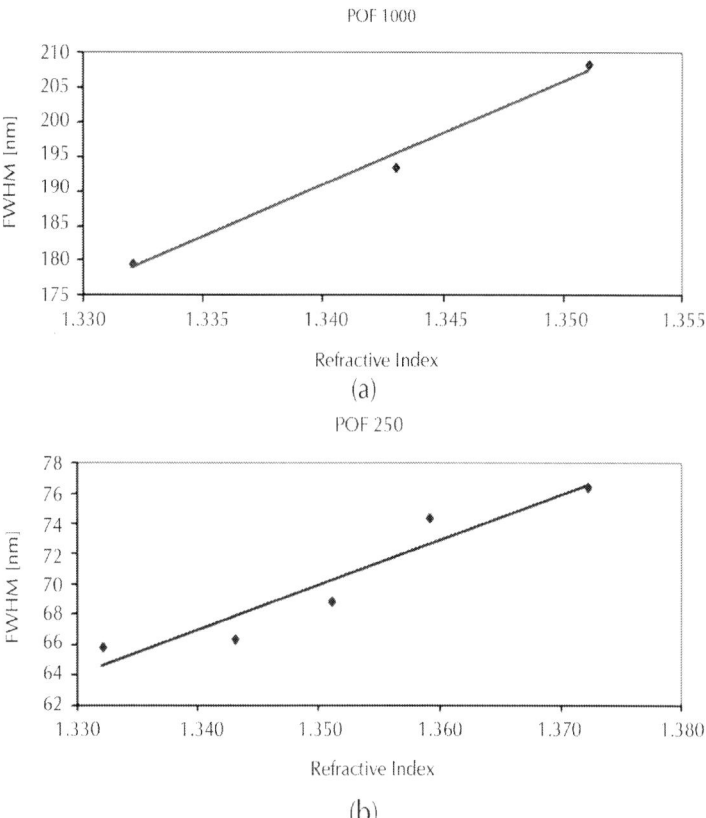

Figure 17: The full width at half maximum of the SPR curve as a function of the refractive index. (a) Configuration with a 1,000 μm diameter POF. (b) Configuration with a 250 μm diameter POF.

The plasmon resonance wavelength as a function of the full width at half maximum of the SPR curve is shown in Figure 18. SNR can be calculated as the angular coefficient of the linear fitting reported inFigure 18. From the above figure, it is clear that the SNR increases when the fiber core diameter decreases. It is important to emphasize that the calculation, from experimental data, of the single values of above parameters has been carried out by employing a first-order approach, while the linear fitting does not imply an actual linear relationship but it is just a way to extrapolate a trend and allow an easy comparison between the two sensor systems.

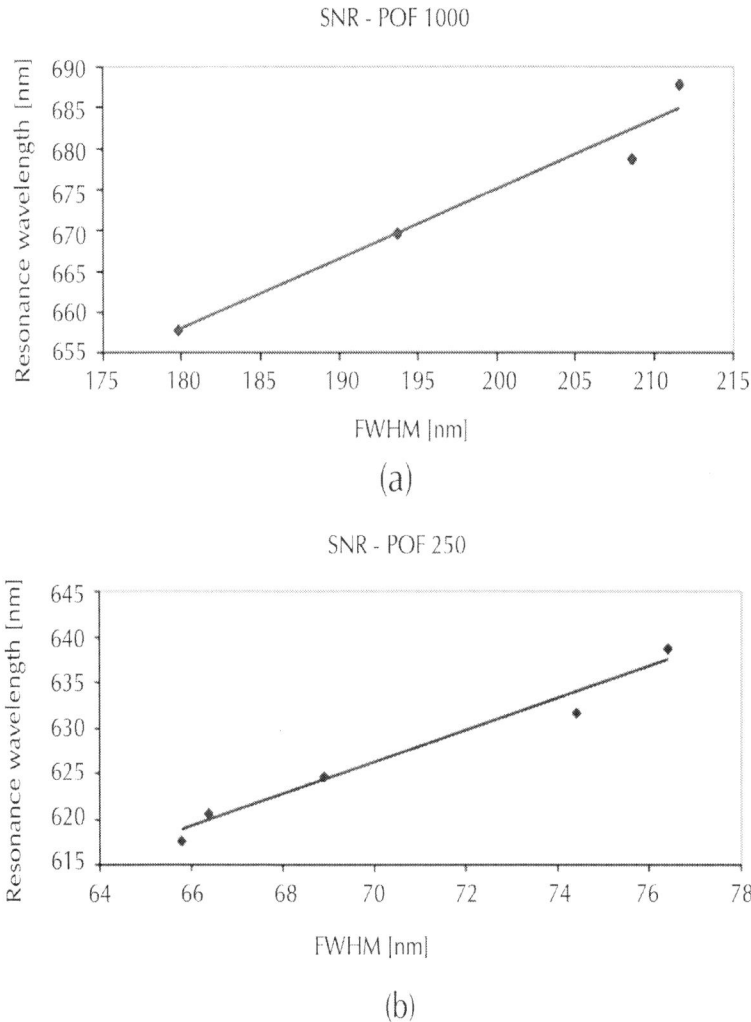

Figure 18: Plasmon resonance wavelength as a function of the full width at half maximum of the SPR curve. (a) Configuration with a 1,000 µm diameter POF. (b) Configuration with a 250 µm diameter POF.

For a clearer comparative analysis between the two sensors with 250 µm and 1,000 µm diameter POFs, Table 2 summarizes the averages values of the experimentally measured performance parameters, evaluated by Matlab software, for external medium refractive index ranging from 1.332 to 1.372.

Table 2: Performance comparison for the two sensors configurations: 250 μm and 1,000 μm diameter POF, respectively

POF Diameter [μm]	Resolution (Δn) [RIU]	Signal-to-noise ratio (SNR)	Sensitivity (S) [nm/RIU]	FWHM/Δn [nm/RIU]
250	0.0027	1.7548	0.549 × 103	0.298 × 103
1,000	0.0010	0.8569	1.325 × 103	1.495 × 103

Sensor Configuration with Taperd POF

Figure 19 shows the optical sensor configuration with a tapered POF. The optical sensor can be realized removing the cladding of a plastic optical fiber along half circumference, heating and stretching it and finally sputtering a thin gold film.

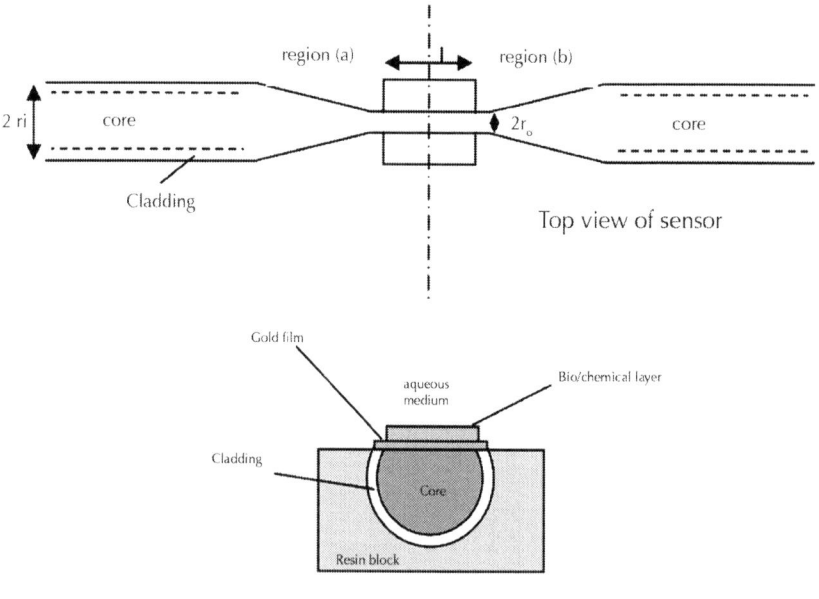

Top view of sensor

Section view of sensor system

Figure 19: Sensor system based on SPR in tapered POF.

The experimental results, presented in this section, are obtained with the following configuration: The plastic optical fiber has a PMMA core of 980 μm and a fluorinated cladding of 20μm. The taper ratio (r_i/r_o) is about 1.5 and the sensing region (L) is about 10 mm in length. The thicknesses of gold layer is about 60 nm. The sensor was realized starting from a plastic optical fiber, without protective jacket, heated (at 150°C) and stretched with a motorized linear positioning stage until the taper ratio reached 1.5. After this step, the POF was embedded in a resin block, and polished with a 5 μm polishing paper in order to remove the cladding and part of the core. After 20 complete strokes following a "8-shaped" pattern in order to completely expose the core, a 1 μm polishing paper was used for another 20 complete strokes with a "8-shaped" pattern. The thin gold film was sputtered by using a sputtering machine (Bal-Tec SCD 500).The sputtering process was repeated three-time with a current of 60 mA for a time of 35 seconds (20 nm for step). On the top of planar gold film it is possible to apply a bio/chemical layer for the selective detection of analytes.

When SPR is achieved in optical fibers, the introduction of the tapered region (a) in fig. 19 helps to reduce the incidence angles of the guided rays in the fiber close to the critical angle of the unclad uniform tapered region. This is obtained by choosing the minimum allowed value of the radius of the output end of the taper so that all the rays remain guided in the uniform core sensing region [9]. After propagating through the uniform region the rays enter the tapered region (b) (see fig. 19) which reconverts the angles of these rays into their initial values so that they can propagate up to the output end of the fiber. Thus, the sensing probe has the minimum diameter such that no ray leaks out and a majority of rays are bound to propagate close to the critical angle, thereby increasing the penetration depth of the evanescent field to almost the maximum value [9].

Figure 20 reports the experimentally obtained SPR transmission spectra, obtained with this tapered POF configuration, normalized to the spectrum achieved with air as the surrounding medium, for six different water-glycerin solutions with refractive index ranging from 1.333 to 1.385.

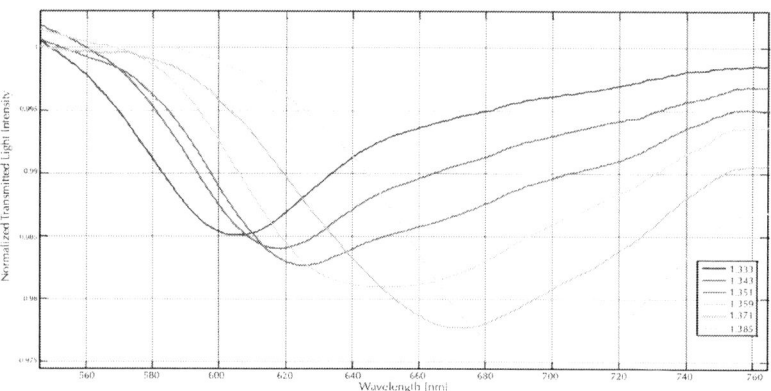

Figure 20: Experimentally obtained SPR transmission spectra, normalized to the air spectrum, for different refractive index of the aqueous medium. Configuration with tapered POF.

Figure 21 shows the resonance wavelength versus the refractive index. In the same figure, it is also presented the linear fitting to the experimental data. The sensitivity, as defined in Equation (2), is the angular coefficient of the linear fitting. Figure 21 shows as the sensitivity increases with the tapered POF configuration.

In this case, i.e. tapered POF without photoresist buffer layer, the sensitivity is about $2*10^3$ (nm/RIU), and it is doubled with respect to the case without tapered POF configuration.

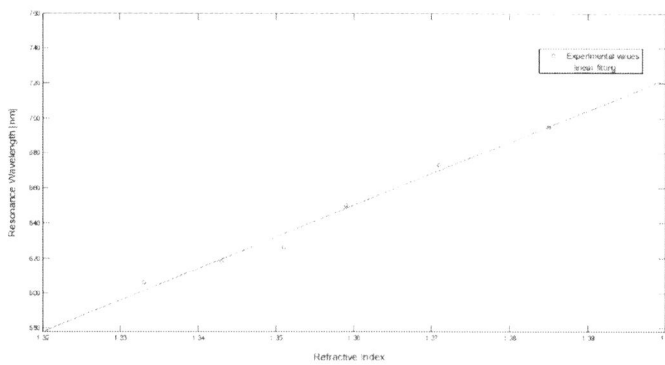

Figure 21: Plasmon resonance wavelength as a function of the refractive index. Configuration with tapered POF.

SPR FOR DETECTION OF BIO/CHEMICAL ANALYTES

When artificial receptors are used for Bio/chemicals detection, the film on the surface of metal selectively recognizes and captures the analyte present in a liquid sample so producing a local increase in the refractive index at the metal surface.

The refractive index change Δn_s induced by the analyte molecules binding to the biorecognition elements can be expressed as [10]:

$$\Delta n_s = \left(\frac{dn}{dc}\right)_{vol} \Delta c_s \qquad (10)$$

where $(dn/dc)_{vol}$ is the volume refractive index increment, and Δc_s is the concentration variation of bound analyte expressed in mass/volume or in any other concentration units in the polymer phase. The value of the refractive index increment depends on the structure of the analyte molecules [11,12].

The refractive index increase gives rise to an increase in the propagation constant of SPW propagating along the metal surface which can be accurately measured, as previously stated.

For this bio-chemical optical sensor with spectral interrogation, the sensitivity is more conveniently defined as:

$$S = \frac{\delta \lambda_{res}}{\delta C} \left[\frac{nm}{M}\right] \qquad (11)$$

In other words, the sensitivity can be defined by calculating the shift in resonance wavelength per unit change in analyte concentration (nm/M).

CONCLUSIONS

In this chapter we have presented an analysis of SPR phenomenon, a POF sensor based on SPR with the related experimental configurations, a performance comparison of sensors based on SPR in POF and a possible implementation as biosensors. The presented devices are based on the excitation of surface plasmons at the interface between an under test medium (aqueous medium) and a thin planar gold layer deposited on a modified plastic optical fiber. Therefore, the proposed sensing head, being low cost and relatively easy to realize, may be very attractive for bio/chemical sensor implementation [6,7].

REFERENCES

1. J. Homola, Present and future of surface plasmon resonance biosensors, Anal. Bioanal. Chem. 377, (2003) 528–539.
2. R.C. Jorgenson, S.S. Yee, A fiber-optic chemical sensor based on surface plasmon resonance, Sens. Actuators B: Chem. 12, (1993) 213–220.
3. M. Kanso, S. Cuenot, G. Louarn, Sensitivity of optical fiber sensor based on surface plasmon resonance: Modeling and experiments, Plasmonics 3, (2008) 49–57.
4. N. Cennamo, D. Massarotti, L. Conte, L. Zeni, Low cost sensors based on SPR in a plastic optical fiber for biosensor implementation, Sensors 11, (2011) 11752–11760.
5. N. Cennamo, D. Massarotti, R. Galatus, L. Conte, L. Zeni, Performance Comparison of Two Sensors Based on Surface Plasmon Resonance in a Plastic Optical Fiber, Sensors 13, (2013) 721-735.
6. N. Cennamo, A. Varriale, A. Pennacchio, M. Staiano, D. Massarotti, L. Zeni, S. D'Auria, An innovative plastic optical fiber-based biosensor for new bio/applications. The Case of Celiac Disease, Sens. Actuators B: Chem. 176, (2013) 1008–1014.
7. N. Cennamo, M. Pesavento, G. D'Agostino, R. Galatus, L. Bibbò, L. Zeni, Detection of trinitrotoluene based on SPR in molecularly imprinted polymer on plastic optical fiber, Proceedings of SPIE 0277-786X, V. 8794, Fifth European Workshop on Optical Fibre Sensors, Kraków, Poland 19-22 May 2013

8. Dwivedi, Y.S.; Sharma, A.K.; Gupta, B.D. Influence of design parameters on the performance of a SPR based fiber optic sensor. Plasmonics 2008, 3, 79–86
9. R. K. Verna, A. K. Sharma, B. D. Gupta, Modeling of Tapered Fiber-Optic Surface Plasmon Resonance Sensor With Enhanced Sensitivity, IEEE Photonics Technology Letters, VOL. 19, NO. 22, (2007) 1786- 1788.
10. J. Homola, Surface Plasmon Resonance Based Sensors, Springer Series on Chemical Sensors and Biosensors, Springer-Verlag, Berlin-Heidelberg-New York, 2006
11. A. Abbas, M. J. Linman, Q. Cheng, New trends in instrumental design for surface plasmon resonance-based biosensors, Biosensors and Bioelectronics 26, (2011) 1815–1824.
12. S. Scarano, M. Mascini, A. P.F. Turner, M. Minunni, Surface plasmon resonance imaging for affinity-based biosensors, Biosensors and Bioelectronics 25, (2010) 957–966.

Chapter 7

A Methodology for Simultaneous Process and Product Design in the Formulated Consumer Products Industry: The Case Study of the Detergent Business

Mariano Martín[a] and Alberto Martínez[b]

[a]Department of Chemical Engineering, University of Salamanca, Plz. Caidos 1-5, 37008, Spain
[b]Procter and Gamble, Brussels Innovation Center, Temselaan 100, 1853 Strombeek-Bever, Belgium

ABSTRACT

In this work we present a mathematical optimization based methodology for simultaneous formulae and process design in the consumer product business applied to the case of the laundry detergent. The design of a new detergent is formulated as a modified pooling problem including process, performance, processability and environmental constraints. This new features add a number of nonlinearities related to the modeling of the different aspects of the process and customer acceptance. The problem becomes a multiobjective optimization problem that is solved using the ε-constraint method with global optimization techniques to minimize of the environmental impact while minimizing the production cost for a couple of case studies. As future work, further process, product and legal constraints can be added to make the problem more realistic.

INTRODUCTION

Over the last two decades, product design has been gaining importance in the field of chemical engineering from two points of view: the design of the product itself and the production process. With the tighter margins, the integrated chemical product and process design has become important for the profitability of any product and the efforts have been placed on better understand and model consumer perception in order to define the properties of the product (Moggridge and Cussler, 2000, Cussler et al., 2010, Uhlemann and Rei, 2009, Bagajewicz et al., 2011 and Teixeira et al., 2012). This fact relies on the specification of a chemical-based product together with the main design parameters corresponding to the manufacturing process. The product/process specification should take into account product functionalities and attributes valued by customers, as well as technical and economical feasibility of its production at a commercial scale. Generic methodologies to guide the solution of such an integrated problem all the way from customer needs to product manufacturing have been proposed. Thus, in the literature we find four main stages to design a product such as consumer needs, ideas generation, selection and manufacture (Cussler et al., 2010, Wibowo and NG, 2001, Wibowo and NG, 2002, Moggridge and Cussler, 2000,Westerberg and Subrahmanian, 2000, Gani, 2004a, Gani, 2004b, Almeida-Rivera et al., 2007, Bongers and Almeida-Rivera, 2009 and Conte et al., 2011). The basic idea behind these methodologies is to drive decisions based on product quality factors related to customer satisfaction, that must be first identified are then translated to a product/process technical specification (Bagajewicz, 2007, Bagajewicz et al., 2011,Smith and Ierapepritou, 2010, Korichi et al., 2008a, Korichi et al., 2008b, Siddhaye et al., 2000 and Bernardo and Saraiva, 2005). It is important to highlight the multidisciplinary nature of this problem where market studies along with an engineering approach are needed to understand the needs and solve the challenges presented by the consumer expectations.

Typical products such as pharmaceuticals, cosmetics, food products, perfumes, softeners or detergents are defined as "performance products" whose value depends on that performance. The need for these kinds of products depends on their use. Many times the function of the product can be related to a physical or chemical phenomena, such as action on a certain illness or detergency, some others the need depends on a particular costumer attribute such as good taste or smell, nice feeling of the fabric in which case the design problem becomes harder since there is no complete understanding of the human perception (Moggridge and Cussler, 2000, Cussler et al., 2010, Uhlemann and Rei, 2009 and Teixeira et al., 2012). The problem can even become more difficult to handle when cultural factors come into play because sociological studies are also required.

In this paper we present a methodology for the simultaneous process and product design applied to performance products. The paper is organized as follows. In Section 2 we present the problem formulation and different products whose production can be analyzed using this approach. Section 3 presents our case study related to the production process of powder detergent, the main ingredients and the use as well as the methods to come up with models for the various constraints. In Section 4 we describe the modeling approach including typical pooling problem constrains (Misener et al., 2010, Misener et al., 2011, Lee and Grossmann, 2003, Ruiz and Grossmann, 2010, Karuppiah and Grossmann, 2006 and Karuppiah et al., 2008), together with the process feasibility and performance models for the detergent. Finally, environmental constraints are also added creating a multiobjective optimization problem. In Section 5 we present the results using some case studies and finally we discuss some conclusions in Section 6.

GENERAL PROBLEM FORMULATION

Most of the performance products are based on the proper formulation of a number of ingredients which provide certain properties we need in our final product. For instance, pharmaceutical products or drugs that, apart from the active principle, include excipients, solvents, etc., Nguyen et al. (2011) and Phan et al. (2011). Another example is the food industry, so that the products have the proper taste and flavor, characterized by oxidative stability, textural properties given by crystal habit, solidification point and conservants so that the product last longer, Avramenko and Kraslawski (2008) and Almeida-Rivera's et al. (2007) or the cleaning products industry including shampoos, liquid detergents (Lai, 2005), dish washing or dry laundry (Lai, 2005, Bayly et al., 2006, Showell, 2006 and Smulders et al., 2007) or tooth pastes to provide cleaning of different surfaces, body creams and lotions so that they are smooth to the skin, not greasy, with the proper thickness, usually related to lotions efficiency, stability and its acceptability by the consumer, creaminess, spreadability, with the proper absorption rate on the skin (Bagajewicz et al., 2011). The methodology we present here somehow involves Almeida-Rivera's et al. (2007) methodology on determining process and product characteristics together with Bagajewicz's et al. (2011) work on evaluating the consumer acceptance together with the environmental constraints over imposed on the pooling problem based on Misener's et al. (2010) paper with the addition of modeling process constraints related to the production or processability of the mixture for the simultaneous process and product optimization aiming minimum production cost with the lower environmental impact for families of products.

The first stage consists of identifying the main ingredients that are required. Traditionally formulation books are available for many products such as detergents (Showell, 2006 and Smulders et al., 2007), drugs (Niazi, 2009) creams and lotions (Flick, 2001), but industrial experience is key at this stage.

Next we need to understand the effect that the level of the ingredients has on the need to be satisfied, either good taste and flavor, cleaning performance, effect on the particular illness. These constraints are hard to obtain since a number of experiments on consumer response must be done to evaluate these variables (Bagajewicz et al., 2011, Cireli et al., 2004 and Teixeira et al., 2012). In some cases there is a physical process behind that can be used as reference such as stain removal in the case of detergents (Cireli et al., 2004), the spreadability, the thickness or the smoothness of a lotion (Bagajewicz et al., 2011) or the fragrance that a perfume leaves on the textile (Teixeira et al., 2012) and the processability of the mixture (Khalloufi et al., 2010 and Khalloufi et al., 2011) in which cases the model development is based on physical and chemical principles. However, in most cases the formulation is too complex and involves too many ingredients to determine such relationships from basic chemical or physical principles and other techniques based on experimental or consumer tests and statistical analysis are used to develop mathematical models that relate the ingredients to the performance of the products (Nguyen et al., 2011 and Phan et al., 2011).

The third stage consists of processing the mixture from the basic ingredients to the final product. A number of chemical and physical transformations are involved. Typically we have blending (Misener et al., 2011) of the ingredients to determine the composition. The actual mix must be processed and the rheological behavior and physical properties of the mixture determines the feasibility of the mixing or pumping. In some cases, we also have particle drying, which depends on the particle size, the air flow and temperature, tablet formation, etc., and the properties of the final product.

For modeling stages two and three, it is possible to use any of the methods commented in Martín and Grossmann (2012) since these processes share a lot of characteristics, from the mathematical modeling point of view, with the bioprocesses commented in that paper. In general there is lack of physical understanding of the some of the processes, together with high complexity, large number of variables and few data available in the literature. Thus in Table 1 we summarize the methods that can be used to obtain mathematical models for the process and performance constraints.

Table 1: Methods for process and performance constraints modeling

Method	Basis	Advantages	Disadvantages
Short cut models[a,b]	Basic first principles	Simple representation	Difficulties in modeling non ideal or complex behavior
Dimensionless correlations[c,d]	Pi theorem Grouping the variables into dimensionless numbers	The models have physical meaning and scale-up issues are usually accounted for.	The physics may not be fully captured if some of the variables are not included.
Rules of thumb models[d,e,f,g]	Experience. Experimental Know how	Well tested rules	Operational data are scarce. Their use is limited to the validated range
Factorial design of experiments (DOE)[h]	Statistical analysis	They allow the systematic study of the effect of a large number of variables	They can only predict within the range they have been obtained for. The models are subjected to scale up problems. Tight bounds for the variables are essential
Empirical correlations[i,j]	Experimental studies	The models involve physical meaning and experimental background	They are only valid for the range of experimental conditions. Sometimes the relation-ship between variables may not be easy to find when several variables are involved
Mechanistics models[k]	Physical principles	Modeling based on first principles	Complex models
Neural networks[l]	Statistical analysis	Good fitting and prediction when trained	Require a lot of sampling
Kringing[m]	Interpolation	Good representation of complex surfaces of response	Require proper sampling

[a]Douglas (1988).
[b]Biegler et al. (1997).
[c]Buckingham (1914).
[d]Branan (2000).
[e]Sinnott (1999).
[f]Wallas (1990).
[g]Perry and Green (1997).

[h]Montgomery (2001).
[i]Phillips et al. (2007).
[j]Martín and Grossmann (2011a).
[k]Martín and Grossmann (2011b).
[l]Henao and Maravelias (2011).
[m]Caballero and Grossmann (2008).

Furthermore, the final product needs to meet certain laws either environmental or public health ones, which most of the times are related to the composition (Smulders et al., 2007). For this purpose some of the waste streams are required to be reused, and reblend is a common practice in many cases.

Finally, in any case we are interested in the sustainable production of the final product which includes environmental burden to be minimized at the minimum cost. Therefore we have a multiobjective optimization problem with non convex terms, mainly coming from modeling the process, performance and blending constraints, which require the use of global optimization techniques for its solution.

CASE STUDY OF THE LAUNDRY BUSINESS

Detergents are one of these consumer products which play very important roles in our daily lives for personal, household surface and fabric care (Bozetine et al., 2008). The detergent ingredients market is huge. In 1996, US demand was 3 billion kilograms while sales were $7 billion (Showell, 2006) while worldwide total textile washing products amounted to sales of $ 17.7×10^9 in the seven major single markets (USA, Japan, Germany, UK, France, Italy, and Spain) in 1999 (Smulders et al., 2007). The highly competitive market and the reduced margin of benefits in the laundry business create opportunities for systematic product design of the detergent.

The detergency of the product depends on its formulation including a large number of chemicals, as we will discuss later, whose effect on the performance of the detergent and the chemical stability and processability of the formula determines its composition, considering that all other factors such as surface to be cleaned, soil, water hardness, and temperature are fixed (Cireli et al., 2004). Furthermore, what the costumer understands as clean and fresh also depends on the specific market which is the target of the product (Bayly et al., 2006, Showell, 2006, Smulders et al., 2007 and Boerefijn et al., 2007). The detergent formulation is a complex blend of over 95% of powders: inorganic salts of alkaline metal called builders, and a system of surfactants composed of anionic and nonionic organic molecules. We have to bear in mind that not all products contain all ingredients (Showell, 2006, Smulders et al., 2007 and Boerefijn et al., 2007). While in the last decade innovations such as compacts and unit-dose systems were introduced (Unit-dose systems were introduced in 1998. Powder tablets were launched in 1998–1999 and liquid unit-dose – i.e., liquid detergent containing pouches that dissolve when immersed in water–were introduced in 2001.), regular powder formulations have still important market shares in many European countries (Van Hoof et al., 2003). Worldwide, powder detergents continue to constitute the largest section of the total textile washing and cleaning products market, both in volume and value terms. Exceptions are the USA, where liquid laundry products took over the lead in the 1990s; Indonesia, whose market is dominated by detergent pastes; and the Philippines, where the vast majority of the consumers use soap bars. India is the world's largest single market by volume for soap bars, but even though, powders dominate and continue to grow at the expense of bars.

In this paper we combine blending operations, with process, performance and environmental burden constraints (Saouter and van Hoof, 2002) for the optimization of the formulation of detergents. Besides good washing performances, processability constraints such as particle size, cake strength together with reblending so that we reuse wastes from previous operations is also considered with no environmental burden but also economic benefit. Only quantitative aspects of the formulation, on a physical–chemical basis, were considered, although psychological aspects may also be important regarding the market for the product and can be included as constraints as long as there is mathematical evidence of them.

Tipical Components of a Detergent Formulation

Out of the long list of ingredients typically used in detergent formulation we are going to classify them into eleven groups (cleaning101, 2012, Dall'Acqua et al., 1999, Smulders et al., 2007 and Boerefijn et al., 2007).

Surfactants/cleaning agents: Their task is to Improve the wetting ability of water, loosen and remove soil with the aid of wash action, then emulsify, solubilize, or suspend soils in the wash solution until soils are washed away.

Builders: Their function is to bind the calcium and magnesium ions coming mainly from the water and in part from the dirt or the textiles to enhance or "build" the cleaning efficiency of the surfactant by inactivating water hardness minerals.

Enzymes: Their main effect is to break down the peptide bonds by enzymatic hydrolysis of complex stains and soils, including protein-based stains (grass and blood) and starch-based stains common to many foods. Enzymes can also improve the appearance and feel of fabrics.

Polymers: They help to capture and hold soils and dyes, sending them down the drain to avoid re-depositing on washed fabrics.

Bleach: They docolor oxidatively the stains that have an affinity to fibre and cannot be washed out through degradation of the chromophore systems. We can distinguish two main oxidative processes, peroxide (typical in Europe) and hypochlorite based one respectively.

Softeners: Reduce fabric friction or static electricity, and help to provide a soft, fluffy appearance for fabrics.

Stabilizers: Maintain high-sudsing function, where suds level is an important indicator of cleaning power. They also help maintain stability of the product and its shelf life, especially the enzymes and oxygen bleach.

Preservatives: Substances used to protect against natural effects of product aging, e.g., decay, discoloration, oxidation, and bacterial attack. They can also protect color and fragrance.

Solvents: Prevent separation or deterioration of ingredients in liquid products.

Fragrances: Provide pleasant and fresh scent to fabrics and cover the odors of the alkaline wash liquor generated in the washing solution.

Colorants: Added to lend individuality to the product, or dramatize a special additive contributing to product performance. Favorite colors are pink, blue and green

Although large in volume – see Table 2 – the sales value of builders (22%) is dwarfed by surfactants (45%) and additives for secondary benefits (33%), Smulders et al. (1997).

Table 2: Composition of powdered laundry detergent formulations in Europe (1997) Smulders et al. (1997)

Ingredient	Weight % (traditional)	Weight % (compact)
Surfactants	10–15	10–25
Builders	28–55	28–48
Bleach	10–25	10–20
Bleach activator	1–3	3–8
Fillers	5–30	None
Corrosion inhibitors	2–6	2–6
Enzymes	0.3–0.8	0.5–2.0
Optical whitening agents	0.1–0.3	0.1–0.3
Anti-foam agents	0.1–4.0	0.1–2.0
Water	To 100%	To 100%

Production Process

Detergent production follows a number of stages which have an important contribution to the required formulae. The process comprises mixing, drying and densification, though the order can change from company to company. In addition to this, there may be granular or spray-on admix components, e.g., enzyme, perfume and bleach

(Boerefijn et al., 2007). Here we consider the flowsheet presented in Fig. 1that consists of 6 stages for the sake of the example such as (1) slurry making, (2) pumping, (3) atomization, (4) drying, (5) cooling and classification and (6) finishing (Showell, 2006, Bayly et al., 2006,ktron, 2012, Boerefijn et al., 2007, Smulders et al., 2007 and Boerefijn et al., 2007).

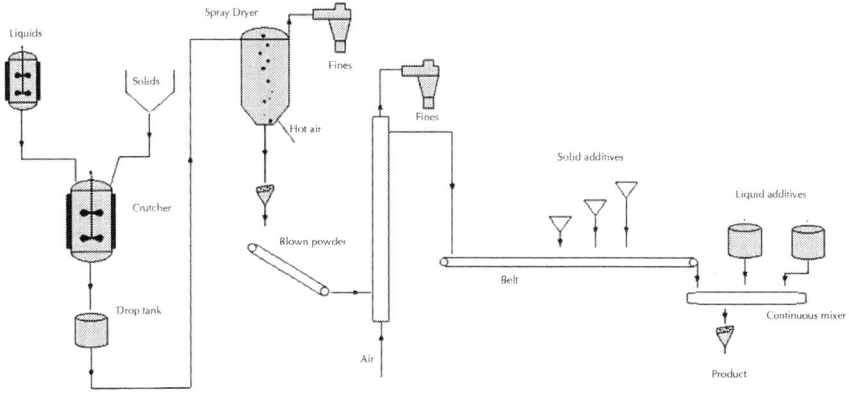

Figure 1: Detergent production process.

Slurry making: The first stage consists of preparing a homogeneous slurry with the minimum water content, since a drying stage will be required later. The challenge is mixing power and liquid raw materials in a tank known as crutcher. From the crutcher the mixture is sent to another tank, the drop tank, where further mixing occurs and some chemical and physical processes such as phase changes and crystallization take place. (Showell, 2006, Bayly et al., 2006, Smulders et al., 2007 and Boerefijn et al., 2007)

Pumping: Due to the low water content inorganic lumps can easily be generated blocking the nozzles of the spray drier. Thus the mixture is usually pumped through a filter and a disintegrator to break up any such lumps before passing to the main slurry pump which is the one used to reach up to 100 bar for the atomization process (Showell, 2006, Bayly et al., 2006 and Smulders et al., 2007).

Atomization: In order to dry, the slurry is atomized to generate small drops. The typical equipment is a spray dryer with a number of nozzles operating at high pressure and distributed at different levels along the column and at a distance from the wall to avoid build up. Thus smaller towers typically run with smaller nozzles and to reduce residence time. Two problems are often encountered with spray nozzles: nozzle wear (due to the abrasion of the slurry) and nozzle blockages (sometimes due to agglomeration) (Showell, 2006, Bayly et al., 2006 and Smulders et al., 2007).

Drying: Countercurrent spray drying towers are used to dry the droplets and operate with an inlet temperature of 300 °C. The cylindrical section of the diameter is typically in the range 3–10 m with geometry such that it improves air and thus temperature distribution, preventing fires and bad product quality. The air can enter radially or with a swirl being this last one preferred many times since stabilizes the airflow and reduces the exhaust temperature. With the exhaust air, fines are carried over and must be recovered using cyclones or filters (Showell, 2006, Bayly et al., 2006, Smulders et al., 2007 and Boerefijn et al., 2007). This will be one of the sources of wastes we can reuse.

Cooling and classification: The powder temperature exiting the tower base is typically over 70 °C. Thus it has to be cooled down to allow the addition of temperature-sensitive additives such as enzymes to be mixed in. Typically, this is done in an airlift, which transports the powder to the top of the building. A gravity separator is then typically used to disengage the particles from the air. This type of system also performs an initial classification of the powder as large lumps are not transported up the airlift and fine particles do not disengage from the gravity separator. (These fine particles are subsequently removed by a bag filter.) Further classification is often required to remove coarser particles. This is typically done using mechanical screens. In the majority of granular detergent formulations, the physical properties of the products are determined by the blown powder properties. The key properties are density, particle size distribution, cake strength–flow properties and solubility in water. We briefly comment on them since they will be our process constraints.

Density: Density is an important quality item since most consumers dose by volume, and therefore, the bulk density is the prime variable to control. The typical values are in the range of 250–550 kg/m^3 and depend on many variables such as the formulation (the water content increases the density), slurry air content, process conditions during drying (air temperature increases the density) since they can be broken due to the expansion of air inside them or become agglomerated is not properly dried (Showell, 2006, Bayly et al., 2006, Ebihara and Watano, 2003 and Smulders et al., 2007).

Particle size distribution: A typical volume-based median particle size is 400–500 μm. The spread of the distribution is typically wide with some significant fine and coarse powder resulting from the agglomeration process, which takes place within the tower as wet drops collide. Acceptable ends of the spectrum typically range from <5% below 100 μm to <2% above 800 μm. The former is for dustiness of the powder and subsequent dispersion when wet. The latter is to control appearance and dissolution of the powder over time during the washing process (Showell, 2006, Bayly et al., 2006, Ebihara and Watano, 2003 and Smulders et al., 2007).

Cake strength: The cake strength is often used as a measure of flowability and granule stickiness. These are important properties for both postprocessing and consumer acceptance. The cake strength tends to be an intrinsic function of the formulation and blown powder moisture while the process conditions tend to have only a minor effect (Showell, 2006, Bayly et al., 2006 and Ebihara and Watano, 2003).

Solubility: The consumer prefers a fully soluble granule, therefore insoluble residues left by a granule are undesirable. The key factor for solubility tends to be formulation and component interactions, although process conditions can also play a role. Some of the better-known interactions that can cause insolubles involve sodium silicate—a common ingredient. Poor mixing control with acidic ingredients can result in the production of insoluble silica, and interactions can also occur with zeolite formulations, which will also cause large insolubles (Showell, 2006 and Bayly et al., 2006). The kind of phosphate also has an effect (Shen, 1968).

Finishing is the final step in the detergent manufacturing process, whereby all the constituent parts of the detergent product are combined together. This results in the final product that is packed and sold to the consumer. Finishing generally involves the blending together of some or all of the following dry detergent components into a uniform, homogeneous mix (Showell, 2006, Bayly et al., 2006, Smulders et al., 2007 and Boerefijn et al., 2007) including spray dried detergent granules, surfactant agglomerates, bleaches and bleach activators, enzymes, buffers and fillers, specialty particles. In addition, various liquid sprays may be applied to the product such as perfumes, nonionic surfactants and colorants. Liquid sprays need to be sited to avoid adjacent sprays from overlapping. If overlap occurs, it may result in over-wetting of the powder surface leading to caking and poor product characteristics. It is also important to ensure that the spray is directed onto the powder and does not impinge on the drum wall. Where spray is on the drum wall, excess make-up can occur in the drum and balling can occur in the final product. In mixers where multiple sprays are used, the sprays need to be positioned down the drum in a controlled order, with suitable spacing between materials. An *aging* zone should also be included before discharge. Nozzle selection for the liquid sprays is dependent on the nature of the liquid to be sprayed (e.g., its viscosity), the level of liquid required in the final product, how it interacts with the final product, and the overall production rate. For low viscosity liquids (e.g., perfumes, nonionic, and colorant solutions), pressure atomization is normally sufficient. For higher viscosity liquids, it may be necessary to use two fluid, air atomized nozzles to ensure adequate atomization.

Usage of the Detergent

The costumer is going to buy our product, with the best price to quality ratio, and use it in a washing machine where the working conditions, mainly defined by the consumer, depend on the fabric material and the level of stains. Stains are then removed because of the combination between the mechanical energy provided by the mixing and the physicochemistry provided by the detergent. The final performance must meet the consumer demands, which are region dependent in some cases related to brightness of the clothes, freshness, smell, etc. However, the removal of each of the different stains is related to its chemical composition, particles, bleaches, enzymatic, oily and the duty of the detergent is to perform well for all of them within a reasonable degree (Smulders et al., 2007). Commonly empirical models are developed to determine the effect of the composition of the detergent on the removal, for a fixed washing machine (Cireli et al., 2004).

MATHEMATICAL MODELING: PROBLEM FORMULATION

The problem of the optimization of the formulation of a detergent is presented in Fig. 2 where we mix a number of components in order to get the formulations required for various products available in the market. We consider as ingredients surfactants, builders, bleaches, fillers, antifoam, enzymes, polymers and water as basic ingredients grouping the different types in these classes. This mixture, but those finishing ingredients, is fed to the system and processed through the crutcher, see Fig. 1, followed by the spray drier. However, we consider that the water added here as ingredient H is that remaining in the final product.

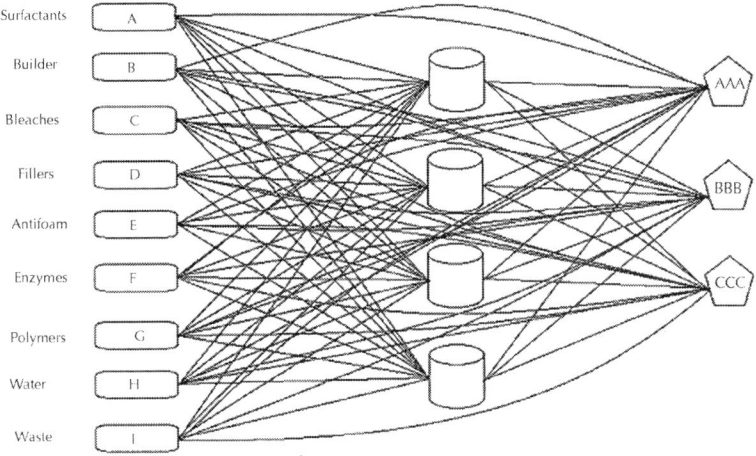

Figure 2: Scheme of feasible mixing processes.

The model has the objective of minimizing the production costs of the set of detergents subjected to typical pooling problem constraints (Misener et al., 2010, Misener et al., 2011, Lee and Grossmann, 2003, Ruiz and Grossmann, 2010 and Karuppiah and Grossmann, 2006), Fig. 2 presents the allowed mixing which includes direct mixing of raw materials to the various final products, generation of intermediates and the final mix between intermediates and initial raw materials to generate the final product.

Min cost production (raw materials)+pools (1)

assuming that the labor, maintenance, and utility costs are similar no matter the product.

S.t.

Feed availability

$$A_i^L \leq \sum_{T_x} x_{i,l} + \sum_{T_z} z_{i,l} \leq A_i^U \quad \forall i \tag{2}$$

Pool capacity

$$\sum_{T_x} x_{i,l} \leq S_l \quad \forall l \tag{3}$$

Product demand

$$D_j^L \leq \sum_{T_y} y_{l,j} + \sum_{T_z} z_{i,j} \leq D_j^U \quad \forall j \tag{4}$$

Material balance

$$\sum_{T_x} x_{i,l} - \sum_{T_y} y_{l,j} \leq 0 \quad \forall l \tag{5}$$

Quality balance

$$\sum_{T_x} C_{i,k} \cdot x_{i,l} - p_{k,l} \sum_{T_y} y_{l,j} \leq 0 \quad \forall l, k \tag{6}$$

Product quality

$$\sum_{T_z} z_{i,j} - \sum_{T_y} p_{k,l} \cdot y_{l,j} \leq P_{j,k}^U \quad \forall l, k$$

$$\sum_{T_z} z_{i,j} - \sum_{T_y} p_{k,l} \cdot y_{l,j} \geq P_{j,k}^L \quad \forall l, k \tag{7}$$

Hard bounds

$$0 \leq x_{i,l} \leq \min\left\{A_i^U, S_l, \sum_{T_y} D_j^U\right\} \quad \forall T_x$$

$$0 \leq y_{j,l} \leq \min\left\{S_l, D_j^U, \sum_{T_x} A_l^U\right\} \quad \forall T_y$$

$$0 \leq z_{j,l} \leq \min\left\{D_j^U, A_l^U\right\} \quad \forall T_z$$

(8)

Furthermore, process, performance and environmental constrains for the different formulae are included:

- *Product performance*: The performance of a particular detergent formulation relies mainly on the ingredients. We are going to develop a model in order to characterize the performance of our formulation based on experimental evidence from the literature but we are not discussing the effect that the washing machine has on cleaning (Brochures, 2012, Yamane and Nakazawa, 1986, Carroll, 1993, Cireli et al., 2004, Lee et al., 2008 and Cussler et al., 2010; Bozetine, 2008). A feasible technique to evaluate the contribution of different variables on the performance of the formulae is a design of experiments (DOE) (Montgomery, 2001). Even though bilinear terms that appear in these kinds of models are non convex, with the proper bounds to the variables, the convex approximations are well known and global optimization algorithms can deal with them easier than if we use other kind of models. Furthermore, the use of DOE is widely spread for predicting an output as function of a number of input variables (Montgomery, 2001, Nguyen et al., 2011 and Phan et al., 2011) without deep evaluation of the physics behind. In Table 3 we present the range of values for the concentrations of the different main components.

Table 3: Range of chemical in the formulation for performance purposes

Surf	Enz	Builder	Polymer	Bleach
15–25	0–0.3	45–85	0–5	0–25

From the 8 components we present as ingredients for the detergent in Fig. 3. We assume that water does not affect detergency, nor does the antifoam or the fillers for the sake of simplicity. The data used for modeling are presented in Table 4. We used a Box-Behnken based model for four of the variables, surfactant, bleach, enzyme and polymer concentration with three levels each, low, medium and high, while the builder adds up to the total 100%. The response, the performance of the mixture, is built assuming that the concentration of surfactants increases detergency, as well as that of the enzymes, the bleaches or the polymers. We assume that this concentration free from the dilutants, fillers, water and antifoam, will determine the performance of the detergent. For processability reasons and also for economic reasons we assume that we need to add those other three components even though dilute the composition. Table 4 presents the level of the different ingredients and the detergency from 1 to 100, which will mean perfect cleaning of the detergent. The table is built in such a way that the more you have of a key component (surfactants, enzymes, polymers and bleaches) the better the performance will be which may not be exactly true in the actual detergent but it will be easy for us and for the sake of the argument since the more of the expensive components the more expensive the detergent and the better the performance will be. Out of those data we can develop a simple multilinear model for the performance of the detergent as function of the main components of the formulation, seen in Eq. (8). Fig. 3 shows the fit between the results in Table 4 and the model production.

$$\text{Performance}(j) = (107 * PQ(j, \text{Surfactants}) + 1872 * PQ(j, \text{Enzyme}) + 53.9 * PQ(j, \text{Builder}) + 134 * PQ(j, \text{Polymer}) + 119 * PQ(j, \text{Bleach})) \qquad (9)$$

A Methodology for Simultaneous Process and Product Design in the... 181

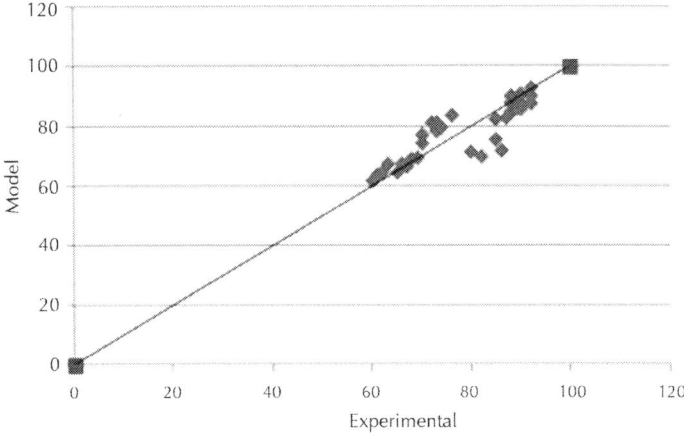

Figure 3: Fit of the model to the experimental results.

Table 4: Assumed data for detergent performance

Surf	Enz	Builder	Polymer	Bleach	Y
15	0	85	0	0	60
15	0	82.5	2.5	0	61
15	0	65	5	15	85
15	0	70	0	15	80
15	0.15	57.35	2.5	25	87
15	0.15	54.85	5	25	88
15	0.15	84.85	0	0	65
15	0.15	82.35	2.5	0	67
15	0.3	64.7	5	15	72
15	0.3	69.7	0	15	70
15	0.3	57.2	2.5	25	90
15	0.3	54.7	5	25	92
20	0	80	0	0	62
20	0	77.5	2.5	0	67
20	0	60	5	15	73
20	0	65	0	15	70
20	0.15	52.35	2.5	25	89

20	0.15	49.85	5	25	88
20	0.15	79.85	0	0	66
20	0.15	77.35	2.5	0	69
20	0.3	59.7	5	15	76
20	0.3	64.7	0	15	74
20	0.3	52.2	2.5	25	88
20	0.3	49.7	5	25	88
25	0	75	0	0	63
25	0	72.5	2.5	0	68
25	0	55	5	15	73
25	0	60	0	15	70
25	0.15	47.35	2.5	25	90
25	0.15	44.85	5	25	92
25	0.15	74.85	0	0	82
25	0.15	72.35	2.5	0	86
25	0.3	54.7	5	15	90
25	0.3	59.7	0	15	85
25	0.3	47.2	2.5	25	90
25	0.3	44.7	5	25	92

- *Environmental constraints*: There are legal regulations to this respect. Thus, in 1989 the NordicCouncil of Ministers decided to introduce a voluntary official ecolabel, which is simple way of communicating environmental work and commitment to customers and suppliers helping in evaluating the impact by considering the content in phosphorous and in fragrances (mainly the allergy effects), the dosage of the detergent, the higher the amount required in a wash the lower the points the formulation gets, the performance and the packing of the detergent (http://www.nordic-ecolabel.org/). These constraints can also be implemented as long as the list of ingredients considered in the formulation is detailed enough.

For the sake of simplicity we are not dealing with specific chemicals but with families of chemicals where a number of different ones are represented, i.e., surfactants. Therefore we are going to define the environmental burden of the final products as given by Eq. (10). We assume that the environmental impact of the production process can be calculated as the weighted sum of the ingredients used where each one has a contribution to the environmental burden. The environmental costs of each of the components involved in the formulation are assumed known based on their nature, $C_{Env,k}$, and are given in Table 5. This table will allow proof of concept and a fair estimation based on the chemical composition of the products, but the actual calculation of the environmental impact is much more difficult mainly because each particular chemical has its own value (Dall'Acqua et al., 1999 and Saouter and van Hoof, 2002) and

equation(10)

$$\begin{aligned}EnvBurden = & \sum_{j}^{n_p} \sum_{i}^{i=indiv.ing} \sum_{k}^{n_{ing}} C_{Env,k} \cdot CC_{i,k} \cdot z_{i,j} \\ & + \sum_{l}^{n_{pools}} \sum_{i}^{i=indiv.ing} \sum_{k}^{n_{ing}} C_{Env,k} \cdot CC_{i,k} \cdot x_{i,l} - \sum_{j}^{n_p} \sum_{wastes}^{} \sum_{k}^{n_{ing}} C_{Env,k} \\ & \cdot CC_{i,k} \cdot z_{i,j} - \sum_{l}^{n_{pools}} \sum_{wastes}^{} \sum_{k}^{n_{ing}} C_{Env,k} \cdot CC_{i,k} \cdot x_{i,l}\end{aligned} \quad (10)$$

Table 5: Weights for environmental function

Ingredient	C_{Env}	Reason
Surfactant	9	The more you have the highest impact
Builder	7	Affects water by binding ions
Bleaches	10	Oxidant
Fillers	5	
Antifoam	5	
Enzymes	2	Reduce operation temperature and dose
Polymers	8	Organic
Water	0	

- *Process constraints*: Among the product physical properties commented above, we are going to focus on cake strength and particle size since there is a bit of evidence reported in the literature in order to come up with models. We are going to use models based on DOE to predict the physical properties due to the convenient form of the mathematical equations used to model the response.

The *particle size* generated at the nozzle of a Spray dryer depends on the air temperature, the flow rate, the air/liquid ratio, the composition of the liquid (liquid viscosity and density) and the pressure (Hayashi et al., 1969 and Cheow et al., 2010; Hecht, 2007). We assume that for constant flow rate at the nozzle and constant temperature, based on Cheow et al. (2009), there is a linear relationship between the water content and the particle size. Moreover, based on the experimental results byHecht et al. (2007) there is a relationship between the amount of carbonates (here considered as fillers for the sake of the argument) and the diameter. Table 6 gathers the values used to develop a correlation, given by Eq. (11). We develop the model assuming a quadratic equation. Fig. 4 shows the fit of Eq. (12) to the data of Table 6. According to the literature, the mean particle size for detergent powder must be in the range between 400 μm and 500 μm (Showell, 2006, Bayly et al., 2006 and Boerefijn et al., 2007). The effect of air flow and operating temperature can be added as long as experimental evidence is available.

Particle(j)=224.5+1509.78*PQ(j,Water)+1000*PQ(j,Filler)−31*PQ(j,Water)*PQ(j,Filler); Particle (j)≥400; Particle (j) ≤500;   (11)

Table 6: Assumed values for modeling the particle size as function of the ingredients

Water (weight fraction)	Fillers (weight fraction)	Particle size (μm)
0.01	0.1	300
0.1	0.1	400
0.2	0.1	500
0.1	0.05	350
0.1	0.125	425
0.1	0.25	550

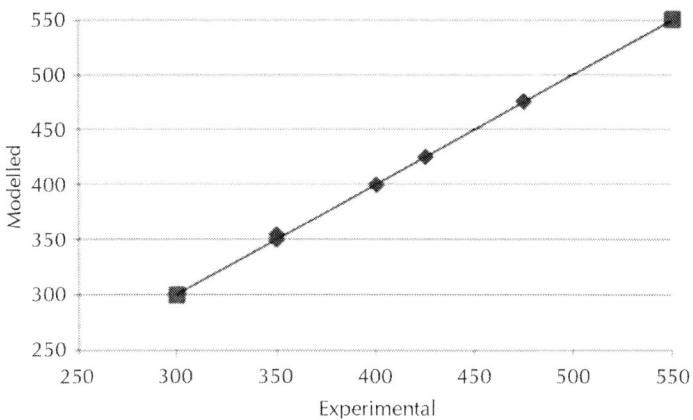

Figure 4: Fit of the particle size (μm) as function of detergent composition.

For the effect of the detergent composition on the *cake strength* the data is scarce. However, in the paper by Ebihara and Watano (2003) there is a comparison between two formulae where the main difference is the polymer content. Cake strength also depends on other factors such as air temperature and flow but no data was found to obtain a model. For sugar powder there are studies on the effect of the water content on the cake strength (Abbas et al., 2010), so we use the effect of the water content presented in that paper together with the one relating the polymer content on the cake strength Ebihara and Watano (2003) to come up with Eq. (12) assuming a quadratic equation as base model. In general a cake strength of around 1 kg is suggested, Ebihara and Watano (2003)

Cakest(j)=2.98*PQ(j,'Water')+2.69*PQ(j,'Polymer')+0.08*PQ(j,'Polymer')*PQ(j,'Water'); Cakest(j)) 0; Cakest(j) 1.4; (12)

We could also include some constraints in order to better meet the aesthetics or the fresh feeling the customer prefers for a particular detergent or softener. However, the lack of data resulting in the fact that it is not possible to come up with realistic mathematical relationships prevent us from pursuing this path any further. Furthermore, at this point we are not going to include the effect that the various cycles of a typical washing machine may have on the cleaning including the length, the temperature and the rotation speed or the effect of the water quality and the different behavior of the various stains and fabric materials (Cireli et al., 2004 and Lee et al., 2008).

CASE STUDIES

For our case studies we consider the production of three products of different quality. A high quality detergent, AAA, medium level detergent, more economic but with lower performance, BBB, and low level detergent, CCC, whose performance is going to be acceptable but for a reduced price. We fix the performance to be higher or equal to a desired level for high medium and low quality product.

Performance(AAA) ≥ 0.95

Performance(BBB) ≥ 0.80

Performance(CCC) ≥ 0.70 (13)

In Appendix A we present the values we have used for the parameters in Eqs. (1), (2), (3), (4), (5), (6), (7),(8) and (9), the base problem, such as raw materials costs and product prices, ranges for the raw material composition (including the composition of the wastes) product composition, availability of raw materials, product demands. We consider up to three different wastes to evaluate their recycling assuming bounds for their availability. The wastes may come from other plants or correspond to the wastes of a previous processing time period. For simplicity, this model considers only one time period and not scheduling issues so that the wastes composition and their availabilities are fixed. We also consider up to five pools, the use of a different performance correlation instead of Eq. (9), more accurate but more complex from the mathematical point of view. The addition of the environmental burden transforms the problem into a multiobjective optimization one. In order to solve it we use the epsilon constraint method that is implemented within the model as Eq. (13),

EnvBurden$\leq \varepsilon$ (14)

as well as process constraints, Eqs. (11) and (12) to the base problem. The wastes are considered to have zero prices to favor their use, while the contribution to the environmental burden is negative since we recycle those materials instead of generating waste streams. BARON 9.0.6 is used to solve the global optimization problem at hand.

Case 1: simple Performance Model

For this example we consider the number of recycled streams to be treated to three while the problem is defined by Eqs. (1), (2), (3), (4), (5), (6), (7), (8), (9), (10), (11) and (12). The presence of the environmental constraint transforms the problem into a multiobjective optimization one where we have production cost and environmental burden as objectives. To solve this problem we are going to use the ε-constraint method to impose constraints to the environmental burden.

Fig. 5 shows the profile of the performance of the optimal product for different maximum environmental burdens, ε. In general the performance of the detergent decreases until it reaches the lower bound we defined for each of them. In this particular case detergent two reaches its lower bound the first of the three at 80%. Pure ingredients and wastes are used to obtain the final products. In Table 7 we present the optimal detergent compositions for this case study. Note that they are the solutions to our case study. In this table we can see which ingredients levels are modified as the environmental burden, ε, is forced to decrease, which gives the idea of the possibilities of this tool in formulating the product. In order to meet the process, performance and environmental constraints the formulation of each of the products evolves. As a rule, the cheapest detergent contains fewer enzymes while fillers and builders decrease as the environmental burden decreases. As a result, the particle size increases, see Fig. 6. On the other hand the cake strength increases as the environmental burden decreases, see Fig. 7, mainly due to the addition of components that keep the performance while reducing the environmental burden, builder content decreases while enzymes and polymers content increase. These changes in the composition of the final product take place as soon as the environmental burden is allowed to be smaller, as we can see in Fig. 6 and Fig. 7.

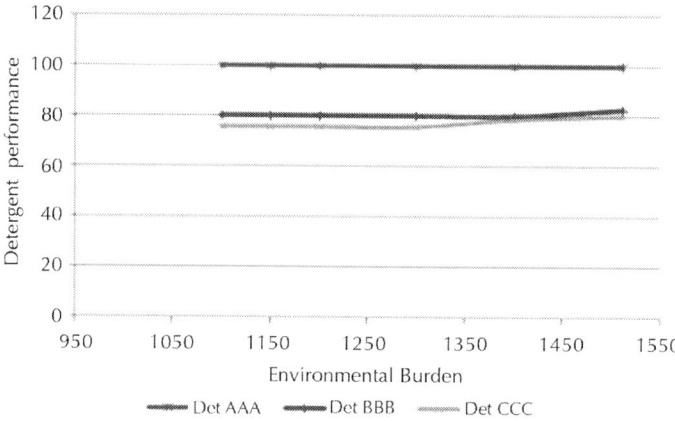

Figure 5: Multiobjective simultaneous process and product optimization.

Table 7: Detergent compositions case study I

	1500			1400			1300			1200			1150			1100		
	AAA	BBB	CCC	AAA	BBB	CCC	AAA	BBB	CCC	AAA	BBB	CCC	AAA	BBB	CCC	AAA	BBB	CCC
Surfactants	0.150	0.150	0.150	0.150	0.150	0.150	0.150	0.150	0.150	0.150	0.150	0.150	0.150	0.150	0.150	0.150	0.150	0.150
Builder	0.501	0.537	0.579	0.501	0.536	0.553	0.467	0.530	0.504	0.467	0.493	0.504	0.467	0.492	0.504	0.467	0.490	0.504
Bleaches	0.050	0.050	0.050	0.050	0.050	0.050	0.050	0.050	0.050	0.050	0.050	0.050	0.050	0.050	0.050	0.050	0.050	0.050
Fillers	0.100	0.119	0.100	0.100	0.115	0.100	0.100	0.109	0.100	0.100	0.111	0.100	0.100	0.113	0.100	0.100	0.114	0.100
Antifoam	0.018	0.001	0.001	0.018	0.001	0.001	0.050	0.005	0.050	0.050	0.027	0.050	0.050	0.028	0.050	0.050	0.028	0.050
Enzymes	0.021	0.010	0.010	0.021	0.010	0.010	0.022	0.010	0.010	0.022	0.010	0.010	0.022	0.010	0.010	0.022	0.010	0.010
Polymers	0.043	0.029	0.020	0.043	0.031	0.020	0.044	0.036	0.020	0.044	0.020	0.020	0.044	0.050	0.020	0.044	0.050	0.020
Water	0.116	0.104	0.090	0.116	0.106	0.116	0.116	0.110	0.116	0.117	0.109	0.116	0.116	0.108	0.116	0.116	0.107	0.116

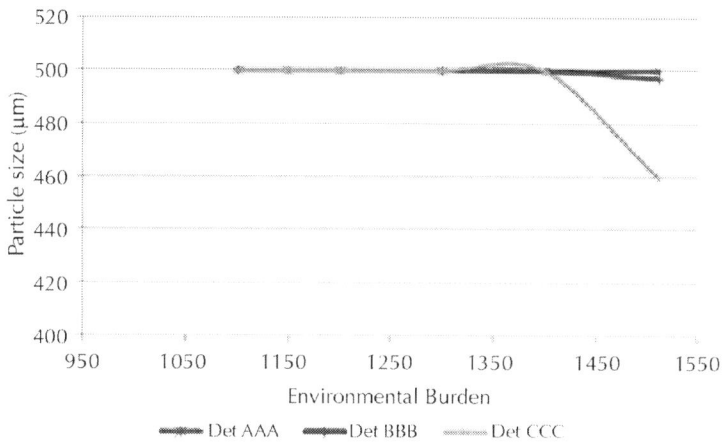

Figure 6: Effect of multiobjective simultaneous process and product optimization on particle size.

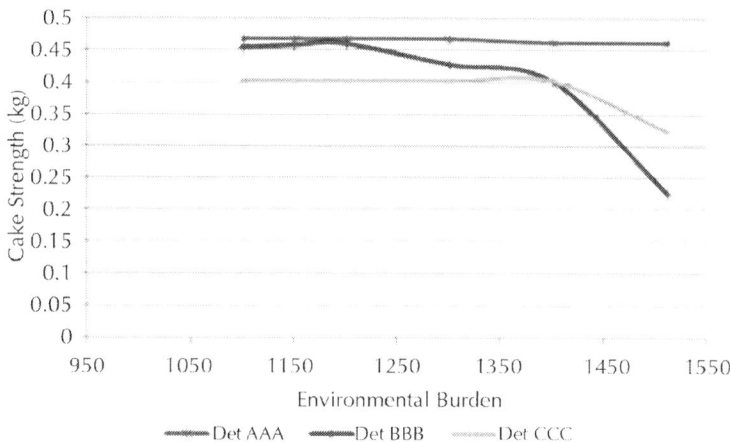

Figure 7: Effect of multiobjective simultaneous process and product optimization on the cake strength.

Fig. 8 presents the pareto frontier for the optimal solutions for the benefit (equal to minus the objective function) at different maximum environmental burdens. As it is expected, the decrease in the environmental burden results in an increase in the production costs or, a decrease in the benefit. However the change in the objective function takes place for values of the environmental burden lower than those for which the particle size and cake strength change. This is representative of the fact that up to 1300 units of environmental burden, the changes in the product composition are focused on maintaining the benefits at the cost of modifying the properties of the products. Below 1300 units, even with composition adjustment to obtain the best cost, we have to sacrifice the benefit to meet all the constraints. The performance decays up to the lower bound for final products BBB and CCC and the changes in the composition modify the properties of the powder as we decrease the environmental burden.

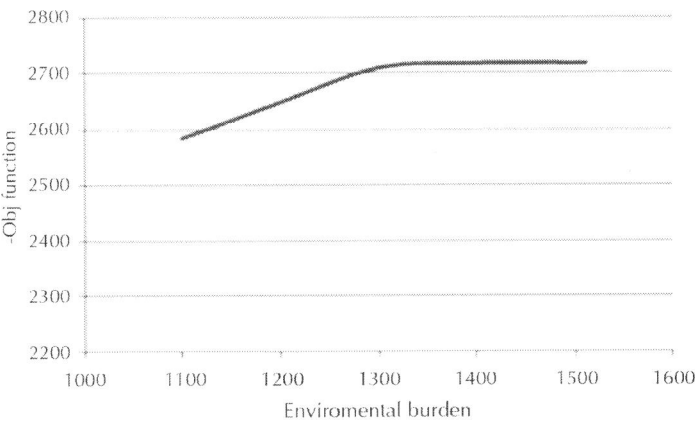

Figure 8: Pareto front.

Case 2: more Accurate Performance Model

For this example we consider the number of recycled streams to be treated to three while the problem is defined by Eqs. (1), (2), (3), (4), (5), (6), (7) and (8) and (10), (11), (12), (13), (14) and (15). Instead of using a linear model for the performance of the detergent, Eq. (9), we look for more accurate models involving bilinear terms given by Eq. (15):

performance(j)=(PQ(j,Surfactant)*(−2040)+PQ(j,Enzyme)*1.26+PQ(j,Builder)*(−0.74)+PQ(j,Polymer)*(−598)+PQ(j,Bleach)*(856)+(747)*PQ(j,Surfactant)*PQ(j,Enzyme)+16.5*PQ(j,Surfactant)*PQ(j,Builder)+27.3*PQ(j,Surfactant)*PQ(j,Polymer)+(−0.07)*PQ(j,Surfactant)*PQ(j,Bleach)+(−92.2)*PQ(j,Enzyme)*PQ(j,Builder)+(15.2)*PQ(j,Enzyme)*PQ(j,Polymer)+PQ(j,Enzyme)*PQ(j,Bleach)*(−511.5)+8.45*PQ(j,Builde r)*PQ(j,Polymer)+(−4.22)*PQ(j,Builder)*PQ(j,Bleach)+(0.0068)*PQ(j, Polymer)*PQ(j,Builder)+(32.1)*PQ(j,Surfactant)^2+(−31.9)*PQ(j,Enzyme)^2+1.15*PQ(j,Builder)^2+(0.058)*PQ(j,Polymer)^2+0.0084*PQ(j,Bleach)^2); (15)

Fig. 9 shows the agreement between the model and the experimental results. As it can be seen the agreement is better than the linear model.

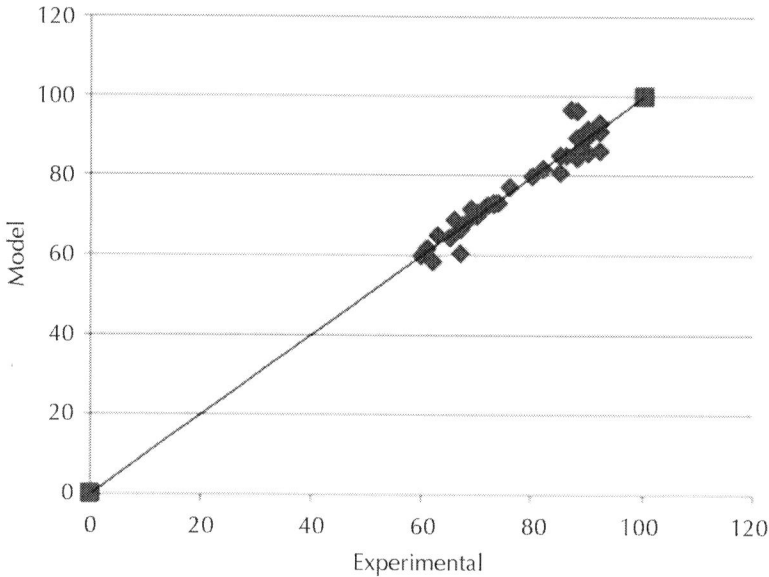

Figure 9: Fit of the non linear model for the performance of the detergent.

When the more accurate model for the performance of the detergent formulation is used, the solution time increases in an order on magnitude. However, the relative small example we have makes its solution feasible within 5–30 min per point in the pareto front. In Fig. 10 we see the profile of optimal solutions for the performance of the detergent as the environmental burden, ε, decreases. While detergents BBB and CCC maintain the performance, detergent AAA performance needs to increase to meet the processability constraints, increasing the production cost and thus lowering the benefits.

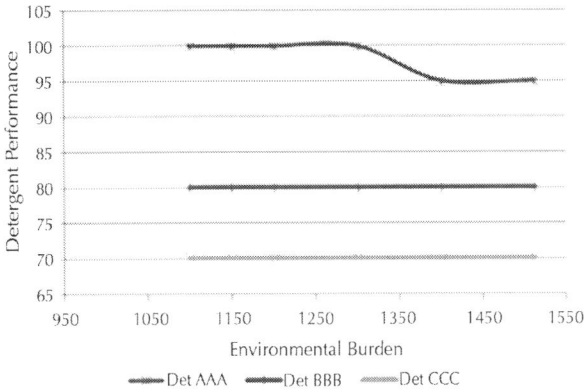

Figure 10: Multiobjective simultaneous process and product optimization.

However, the particle size and the cake strength increases as a result in the changes of the composition of the detergent to reduce the environmental burden, see Fig. 11 and Fig. 12, as in the previous case study. The main change occurs when the environmental burden is reduced to 1300. For that the composition of the detergent is modified to maintain the performance levels while meeting the environmental burden. Table 8 presents the optimal detergent compositions for different environmental burdens. The formulation is slightly different compared to the previous case which highlights the need for good predicting models for product formulation. However, the general rules regarding how the ingredients levels change are similar. Detergent AAA presents higher level of enzymes and, in order to reduce the environmental burden the water and enzyme levels increases while surfactant levels decrease for detergent AAA but decrease for the other two and builder levels decrease for the three of them. Finally, as expected, the decrease in the environmental burden decreases the benefit (minus the objective function which represents the cost) seeFig. 13 for the pareto front. It is important to notice that the figures representing the pareto front are quite similar in both case studies (Fig. 8 and Fig. 13), the production costs are similar for similar environmental burdens. The benefits decrease from environmental burdens of 1300 or below while the composition changes to meet the process constraints as we can see from the fact that the properties of the powder differ at different values for the environmental burden function. Up to this critical value, 1300, the composition can be changed to maintain the benefits with no effect on the benefit, at the cost of the performance. Further on, in order to meet the constraints we sacrifice the production cost. The performance is kept at a constant value; typically the lower bound and the properties of the final product vary as a result of the changes in the composition of the detergent.

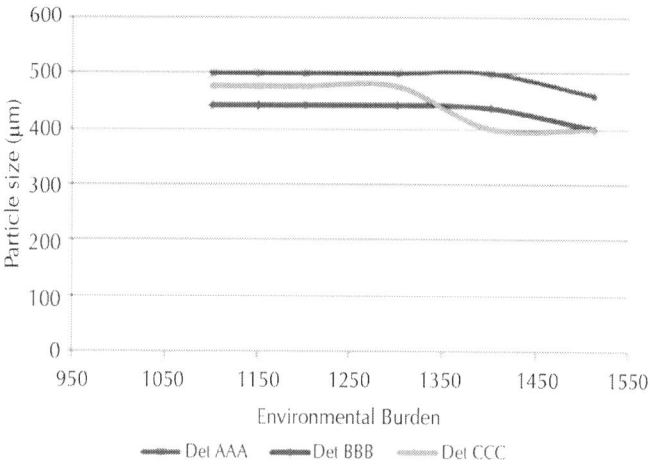

Figure 11: Effect of multiobjective simultaneous process and product optimization on particle size.

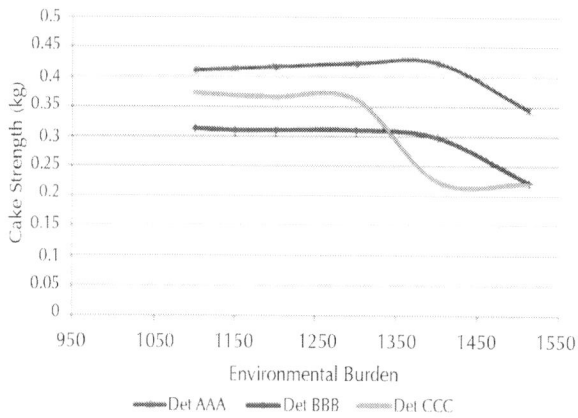

Figure 12: Effect of multiobjective simultaneous process and product optimization on the cake strength.

Table 8: Detergent compositions case study II

	1500			1400			1300			1200			1150			1100		
	AAA	BBB	CCC	AAA	BBB	CCC	AAA	BBB	CCC	AAA	BBB	CCC	AAA	BBB	CCC	AAA	BBB	CCC
Surfactants	0.169	0.234	0.220	0.171	0.248	0.220	0.150	0.250	0.250	0.150	0.250	0.250	0.150	0.250	0.250	0.150	0.250	0.250
Builder	0.481	0.527	0.542	0.459	0.488	0.541	0.438	0.482	0.465	0.437	0.482	0.464	0.437	0.482	0.463	0.436	0.481	0.461
Bleaches	0.050	0.050	0.050	0.050	0.050	0.050	0.050	0.050	0.050	0.050	0.050	0.050	0.050	0.050	0.050	0.050	0.050	0.050
Fillers	0.129	0.100	0.100	0.129	0.100	0.100	0.130	0.100	0.100	0.133	0.100	0.100	0.134	0.100	0.100	0.136	0.100	0.100
Antifoam	0.018	0.001	0.001	0.018	0.001	0.001	0.050	0.001	0.001	0.050	0.001	0.001	0.050	0.001	0.001	0.050	0.001	0.001
Enzymes	0.025	0.010	0.010	0.025	0.010	0.010	0.035	0.010	0.010	0.035	0.010	0.010	0.035	0.010	0.010	0.035	0.010	0.010
Polymers	0.050	0.028	0.027	0.050	0.027	0.028	0.050	0.029	0.023	0.050	0.029	0.024	0.050	0.029	0.025	0.050	0.030	0.027
Water	0.077	0.050	0.050	0.097	0.075	0.050	0.097	0.078	0.101	0.095	0.078	0.101	0.094	0.078	0.101	0.093	0.078	0.101

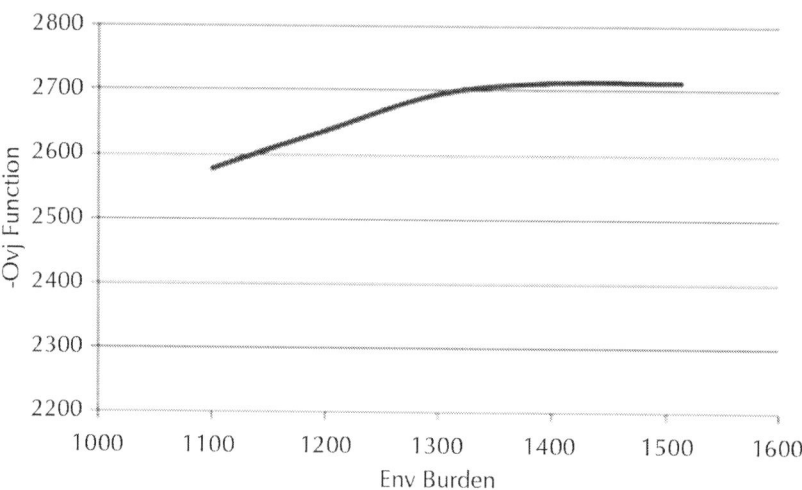

Figure 13: Pareto front.

We have global optimal solutions for each of the problems, which in fact are quit close in terms of cost but due to the differences in the performance function we obtain slightly different compositions for the products. For these two case studies the solution does not use any pool, it is more expensive to use and it does not help. The problem can be enlarged by considering more ingredients (the actual number of ingredients is more that 30) wastes and final products. In that case, we may expect the use of pools become more interesting since the solutions reported no use of the pools even for low cost. However, as a proof of concept and in order to have reasonable control of the differences of the products, wastes and ingredients to produce feasible solutions, since we are making up the data, we have preferred to keep the problem this size with no decrease in the generality of the methodology and formulation of the problem.

CONCLUSIONS AND FUTURE WORK

The use of mathematical programming techniques for designing the optimal formulation of detergents is a powerful technique that allows simultaneously including process, legal and performance constraints to the typical pooling problem constraints with for the design of economical and environmentally friendly formulations.

The more realistic the experimental data we have the more useful the solutions obtained. Thus, process constraints such as slurry viscosity and density (Hayashi et al., 1969; Cheow et al., 2009; Chegini et al., 2008), ingredients mixing and agglomeration constraints or finishing requirements for better customer acceptance such as perfumes or aesthetics (Bayly et al., 2006 and Showell, 2006) can be added which will help design the best product with the lowest cost. Furthermore, psychological aspects may also be important regarding the detergent market and can be included as constraints as long as we can transform the experimental evidence into a proper mathematical formulation.

The complexities in the models that describe the performance or process constraints may result in non global optimal solutions and thus we have to evaluate the best model for the different constraints which has the accuracy needed but which does not add further complexities.

ACKNOWLEDGEMENTS

The authors would like to thank Mr. Zayeed Alam, section head at P&G, for his comments on the manuscript.

REFERENCES

1. Abbas, K.A., Laseken, O., Khalil, S.K., 2010. The significance of glass transition temperature in processing of selected fried food products: a review. Mod. Appl. Sci. 4 (5), 3–21.
2. Almeida-Rivera, C., Jain, P., Bruin, S., Bongers, P., 2007. Integrated product and process design approach for rationalization of food products. Comput. Aided Chem. Eng. 23, 449–454.
3. Avramenko, Y., Kraslawski, A., 2008. Product design: food product formulation. Case Based Des. Appl. Process Eng. 87, 109–116.
4. Bagajewicz, M., 2007. On the role of microeconomics, planning and finances in product design. AIChE J. 53 (12), 3155–3170.
5. Bagajewicz, M., Hill, S., Robben, A., Lopez, H., Sanders, M., Sposato, E., Baade, C., Manora, S., Hey Coradin, J., 2011. Product design in price-competitive markets: a case study of a skin moisturizing lotion. AIChE J. 57 (1), 160–177.

6. Bayly, A.E., Smith, D.J., Roberts, N.S., York, D.W., Capeci, S., 2006. Handbook of detergents Part F: Production. Taylor & Francis Procter & Gamble Company, Cincinnati, OH, USA.
7. Bernardo, F.P., Saraiva, P.M., 2005. Integrated process and product design optimization: a cosmetic emulsion application. Comput. Aided Chem. Eng. 20, 1507–1512.
8. Biegler, L.T., Grossmann, I.E., Westerberg, A.W., 1997. Systematic Methods of Chemical Process Design. Prentice Hall, New Jersey.
9. Bozetine, I., Zaid, T.A., Chitour, C.E., Canselier, J.P., 2008. Optimization of an alkylpolyglucoside-based dishwashing detergent formulation <http://oatao.univ-toulouse.fr/1459/1/Bozetine 1459.pdf>.
10. Boerefijn, R., Dontula, P.R., Kohlus, R., 2007. Detergent Granulation. Handbook of Powder Technology, vol. 11, pp. 673–703.
11. Bongers, P.M.M., Almeida-Rivera, C., 2009. Product driven process synthesis methodology. Comput. Aided Chem. Eng. 26, 231–236.
12. Branan, C.R., 2000. Rules of Thumb for Chemical Engineers, 2nd ed. McGraw Hill. Buckingham, E., 1914. On physically similar systems: illustrations of the use of dimensional equations. Phys. Rev. 4, 345–376.
13. Caballero, J.A., Grossmann, I.E., 2008. An algorithm for the use of surrogate models in modular flowsheet optimization. AIChE J. 54 (10), 2633–2650.
14. Carroll, B.J., 1993. Physical aspects of detergency. Colloids Surf. A: Physicochem. Eng. Aspects 74, 131–167.
15. Chegini, G.R., Khazaei, J., Ghobadian, B., Goudarzi, A.M., 2008. Prediction of process and product parameters in an orange juice spray dryer using artificial neural networks. J. Food Eng. 84, 534–543.
16. Cheow, W.S., Li, S., Hadinoto, K., 2010. Spray drying formulation of hollow spherical aggregates of silica nanoparticles by experimental design. Chem. Eng. Res. Des. 88 (5–6), 673–685.
17. Cireli, A., Sariiṣik, M., Kutlu, B., Yaman, V., 2004. The effects of washing conditions on soil removal in domestic laundering processes. AUTEX Res. J. 4 (2), 101–112.

18. Conte, E., Gani, R., Ng, K.M., 2011. Design of formulations: a systematic methodology. AIChE J. 57, pp. 9, 2431, 2449.
19. Cussler, E.L., Wagner, Q., Maarchal-Heusler, L., 2010. Designing chemical products requires more knowledge of perception. AICHE J. 56 (2), 283–288.
20. Dall'Acqua, S., Fawer, M., Fritschi, R., Allenspach, C., 1999. Life cycle inventories for the production of detergent ingredients. EMPA Report No 224 ISBN 3-905594-09-9.
21. Douglas, J.M., 1988. Conceptual Design of Chemical Processes. McGraw Hill.
22. Ebihara, F., Watano, S., 2003. Development of a novel granular detergent with an interspersion particle comprising an anionic surfactant and a polymeric polycarboxalate. Chem. Pharm. Bull. 51 (6), 743–745.
23. Flick, E.W., 2001. Cosmetic and Toiletry Formulations, 2nd ed. William Andrew Publishing, Noyes, Knovel Library website, 2001. Available at: <http://knovel.com/knovel2/Toc.jsp>.
24. Gani, R., 2004a. Computer-aided methods and tools for chemical product design. Chem. Eng. Res. Des. 82 (A11), 1494–1504.
25. Gani, R., 2004b. Chemical product design: challenges and opportunities. Comput. Chem. Eng. 28, 2441–2457.
26. Hayashi, H., Heldma, N.D.R., Hedrick, T.I., 1969. Influence of spray-drying conditions on size and size distribution of nonfat dry milk particles. J. Dairy Sci. 52 (1), 31–37.
27. Hecht, J.P., Stamper, J.A., Giles, D.K., 2007. Pneumatic Atomization of Laundry Detergent Slurries as affected by Solid Particle Size and Concentration ILASS Americas. In: 20th Annual Conference on Liquid Atomization and Spray Systems, Chicago, IL, May 2007.
28. Henao, C.A., Maravelias, C.T., 2011. Process superstructure optimization using surrogate models. AIChE J. 57 (5), 1216–1232.
29. Karuppiah, R., Grossmann, I.E., 2006. Global optimization for the synthesis of integrated water systems in chemical processes. Comp. Chem. Eng. 30 (4), 650–673.
30. Karuppiah, R., Furman, K.C., Grossmann, I.E., 2008. Global optimization for scheduling refinery crude oil operations. Comput. Chem. Eng. 32 (11), 2745–2766.

31. Khalloufi, S., Almeida-Rivera, C., Janssen, J., van-der-Vaart, M., Bongers, P., 2011. Mathematical model for simulating the springback effect of gel matrixes during drying processes and its experimental validation. Dry. Technol. 29 (16), 1972–1980.
32. Khalloufi, S., Almeida-Rivera, C., Bongers, P., 2010. A fundamental approach and its experimental validation to simulate density as a function of moisture content during drying processes. J. Food Eng. 97 (2), 177–187.
33. Korichi, M., Gerbaud, V., Talou, T., Floquet, P., Meniai, A-H., Nacef, S., 2008a. Computer-aided aroma design. II. Quantitative structure–odour relationship. Chem. Eng. Process.: Process Intensification 47 (11), 1912–1925.
34. Korichi, M., Gerbaud, V., Floquet, P., Meniai, A-H., Nacef, S., Joulia, X., 2008b. Computer aided aroma design I—Molecular knowledge framework. Chem. Eng. Process.: Process Intensification 47 (11), 1902–1911.
35. Lai, K.Y., 2005. Liquid Detergents. CRC Press, Taylor & Francis, Colgate–Palmolive Company, New Jersey, USA.
36. Lee, A., Seo, M.H., Yang, S., Koh, J., Kim, H., 2008. The effects of mechanical actions on washing efficiency. Fibers Polym. 9 (1), 101–106.
37. Lee, S., Grossmann, I.E., 2003. Global optimization of nonlinear generalized disjunctive programming with bilinear equality constraints: applications to process networks. Comput. Chem. Eng. 27, 1557–1575.
38. Martín, M., Grossmann, I.E., 2011a. Energy optimization of Hydrogen production from biomass. Comput. Chem. Eng. 35 (9), 1798–1806.
39. Martín, M., Grossmann, I.E., 2011b. Process optimization of FT—diesel production from biomass. Ind. Eng. Chem. Res. 50 (23), 13485–13499.
40. Martín, M., Grossmann, I.E., 2012. BIOpt: a library of models for optimization of biofuel production processes. Comput. Aided Chem. Eng. 30, 16–20.
41. Misener, R., Thompson, J.P., Floudas, C.A., 2011. APOGEE: global optimization of standard, generalized, and extended pooling problems via linear and logarithmic partitioning schemes. Comput. Chem. Eng. 35, 876–892.

42. Misener, R., Gounaris, C.E., Floudas, C.A., 2010. Mathematical modeling and global optimization of large-scale extended pooling problems with the (EPA) complex emissions constraints. Comput. Chem. Eng. 34, 1432–1456.
43. Moggridge, G.D., Cussler, E.L., 2000. An introduction to chemical product design. Chem. Eng. Res. Des. 78 (1), 5–11.
44. Montgomery, D.C., 2001. Design and Analysis of Experiments. Wiley, NY.
45. Nguyen, D.T., Hoang, M.C., Dang, G.D., 2011. Formulation of Gilanka® capsules containing Gingko biloba standardized extract for single daily administration. HCM City J. Med. Sci. Suppl. 15 (1), 56–60.
46. Phan, N.H., Nguyen, K.V., Dang, G.D., 2011. Formulation of Perindopril erbumine 4mg tablets, using intelligent software as a frame work. HCM City J. Med. Sci. Suppl. 15 (1), 61–65.
47. Niazi, S.K., 2009. Handbook of Pharmaceutical Manufacturing Formulations. CRC Press. Taylor & Francis, Boca Raton, FL.
48. Perry, R.H., Green, D.W., 1997. Chemical Engineer Hand book, 7th ed. McGraw-Hill New, York.
49. Phillips, S., Aden, A., Jechura, J., Dayton, D., Eggeman, T., 2007. Thermochemical Ethanol via Indirect Gasification and Mixed Alcohol Synthesis of Lignocellulosic Biomass. NREL/TP-510-41168.
50. Ruiz, J.P., Grossmann, I.E., 2010. Strengthening of lower bounds in the global optimization of bilinear and concave generalized disjunctive programs. Comput. Chem. Eng. 34, 914–930.
51. Saouter, E., van Hoof, G., 2002. A database for the life cycle assessment of P&G landry detergents. Int. J. LCA 7 (2), 103–114.
52. Siddhaye, S., Camarda, K.V., Topp, E., Southard, M., 2000. Design of novel pharmaceutical products via combinatorial optimization. Comput. Chem. Eng. 24 (2–7), 701–704.
53. Sinnott, R.K., 1999. Coulson and Richardson, Chemical Engineering, vol. 6., 3rd ed. Butterworth Heinemann, Singapur.
54. Shen, C.Y., 1968. Properties of detergent phosphates and their effects on detergent processing. J. Am. Oil Chem. Soc. 45 (7), 510–516.

55. Showell, M.S., 2006. Handbook of Detergents Part D: Formulation. Taylor & Francis Procter & Gamble Company, Cincinnati, OH, USA.
56. Smith, B.V., Ierapepritou, M.G., 2010. Integrative chemical product design strategies: reflecting industry trends and challenges. Comput. Chem. Eng. 34, 857–865.
57. Smulders, E., Krings, P., Verbeek, H., 1997. Recent developments in the field of laundry detergents and cleaning agents. Tenside Surfact. Det. 34, 386.
58. Smulders, E., Rybinski, W., Sung, E., Rahse, W., Steberm, J., Wiebel, M.F., Nordkog, A., 2007. Laundry Detergents. Ullmann's Encyclopedia., http://dx.doi.org/10.1002/14356007.a08 315. pub2.
59. Teixeira, M.A., Rodríguez, O., Rodrigues, S., Martins, I., Rodrigues, A.E., 2012. A case study of product engineering: performance of microencapsulated perfumes on textile applications. AICHE J., http://dx.doi.org/10.1002/aic.12715.
60. Uhlemann, J., Rei, I., 2009. Product design and process engineering using the example of flavors. Chem. Eng. Technol. 33 (2), 199–212.
61. Van Hoof, G., Schowanek, D., Feijtel, T.C.J., 2003. Comparative life-cycle assessment of laundry detergent formulations in the UK. Tenside Surfact. Det. 40 (5), 266–275.
62. Wallas, S.M., 1990. Chemical Process Equipment. Selection and Design. Elsevier, Boston.
63. Westerberg, A.W., Subrahmanian, E., 2000. Product design. Comput. Chem. Eng. 24, 959–966.
64. Wibowo, C., NG, K.M., 2002. Product-centered processing: chemical-based consumer product manufacture. AIChE J. 48, 1212–1230.
65. Wibowo, C., NG, K.M., 2001. Product-oriented process synthesis and development: creams and pastes. AIChE J. 47, 2746–2767.
66. Yamane, I., Nakazawa, T., 1986. Development of zeolite for non phosphate detergents in Japan. Pure Appl. Chem. 58 (19), 1397–1404. <http://www.ktron.com/industries served/Chemical/Laundry Detergent Production.cfm< (last accessed February

2012). <www.cleaning101.com> (last accessed February 2012). <http://online1.ispcorp.com/Brochures/Performance%20Chemicals/hiform.pdf> (last accessed February 2012). <http://www.nordic-ecolabel.org/> (last accessed March 2012).

Chapter 8

Advances in Pressure Swing Adsorption for Gas Separation

Carlos A. Grande

Department of Process Chemistry, SINTEF Materials and Chemistry, Blindern, 0314 Oslo, Norway

ABSTRACT

Pressure swing adsorption (PSA) is a well-established gas separation technique in air separation, gas drying, and hydrogen purification separation. Recently, PSA technology has been applied in other areas like methane purification from natural and biogas and has a tremendous potential to expand its utilization. It is known that the adsorbent material employed in a PSA process is extremely important in defining its properties, but it has also been demonstrated that process engineering can improve the performance of PSA units significantly. This paper aims to provide an overview of the fundamentals of PSA process while focusing specifically on different innovative engineering approaches that contributed to continuous improvement of PSA performance.

INTRODUCTION

Adsorption is the name of the spontaneous phenomenon of attraction that a molecule from a fluid phase experiences when it is close to the surface of a solid, named adsorbent. There are several pristine works that explain this phenomenon in detail [1–18]. Adsorbents are porous solids, preferably having a large surface area per unit mass. Since different molecules have different interactions with the surface of the adsorbent, it is eventually possible to separate them. When the adsorbent is put in contact with a fluid phase, an equilibrium state is achieved after a certain time. This equilibrium establishes the thermodynamic limit of the adsorbent loading for a given fluid phase composition, temperature, and pressure [3]. Information about the adsorption equilibrium of the different species is vital to design and model adsorption processes [19–27]. The time required to achieve the equilibrium state may be also important, particularly when the size of the pores of the adsorbent are close to the size of the molecules to be separated [28–43].

In an adsorption process, the adsorbent used is normally shaped into spherical pellets or extruded. Alternatively, it can be shaped into honeycomb monolithic structures resulting in reduced pressure drop of the system [44–54]. The feed stream is put into contact with the adsorbent that is normally packed in fixed beds. The less adsorbed (light) component will break through the column faster than the other(s). In order to achieve separation, before the other (heavy) component(s) breaks through the column, the feed should be stopped and the adsorbent should be regenerated by desorbing the heavy compound. Since the adsorption equilibrium is given by specific operating conditions (composition, T and P), by changing one of these process parameters it is possible to regenerate the adsorbent.

When the regeneration of the adsorbent is performed by reducing the total pressure of the system, the process is termed pressure swing adsorption (PSA), the total pressure of the system "swings" between high pressure in feed and low pressure in regeneration [55, 56]. The concept was patented in 1932, but its first application was presented thirty years later [57].

Over the years it has been demonstrated that PSA technology can be used in a large variety of applications: hydrogen purification [58–72], air separation [57, 73–80], OBOGS (on-board gas generation system) [81], CO_2 removal [82–84], noble gases (He, Xe, Ar) purification [85–87], CH_4 upgrading [31, 34, 37, 40, 42,88–96], n-iso paraffin separation [5, 97–99], and so forth. The PSA processes are normally associated to low energy consumption when compared to other technologies [12, 55, 100–102].

As a rule of thumb, pressure swing adsorption is preferred to other processes when the concentration of the components to be removed is quite important (more than a few per cent). In such conditions, loading the column with the heavy component is accomplished quite fast and since the pressure of the system can be changed rapidly, the time between adsorption and regeneration is balanced. When the concentration is low, the adsorption step may take much longer and other options like temperature swing adsorption (TSA) can be considered [12].

The behaviour of the PSA unit is mainly determined by the adsorbent employed for the separation. However, the engineering of the PSA unit is also an important aspect. In fact, the main task of defining a PSA unit is to select correctly the adsorbent to be employed [103]. After that, all the engineering efforts should be placed in defining an effective strategy to regenerate the adsorbent. Thus, the advances obtained in PSA units can be divided in two main domains: the discovery of new adsorbents (material science) and new and more efficient ways to use and regenerate the adsorbent (engineering).

This work provides an overview of PSA processes and its evolution on time. The most important industrial applications of PSA processes will be used to address its technological evolution: air separation and hydrogen purification. A growing market of PSA, CH_4-CO_2 separation, will also be used for some specific examples. Although it is not intended to describe the state-of-the-art of materials science, an example of the effect of different adsorbent materials in PSA operation will be provided. Finally, the effect of different regeneration protocols and the reduction of the overall cycle time (Rapid Pressure Swing Adsorption) are discussed.

FUNDAMENTALS OF PRESSURE SWING ADSORPTION

The essential feature of the PSA is that when the adsorbent is saturated, using a sequential valve arrangement, the feed is stopped and simultaneously the total pressure of the column is reduced. The reduction in pressure results in a partial desorption of all the species loaded in the column, "regenerating" the adsorbent. Since this process was patented after TSA, it was initially known as "heatless" process. The first patent application where PSA technology was described, was presented by Charles Skarstrom for oxygen enrichment [57]. A scheme of the two-column PSA introduced in that patent is shown in Figure 1. In order to operate such unit cyclically, a column experiences a series of "steps": events like opening and closing valves and changing flowrate direction for example. The sum of all the steps is termed as "cycle". Even when the process is unsteady, after some cycles it reaches a Cyclic Steady State, CSS. When CSS is achieved, the performance of the cycles of the PSA is constant over time. It should be noted that since this process sometimes involve substantial amount of heat generation, there can be multiple CSS [104].

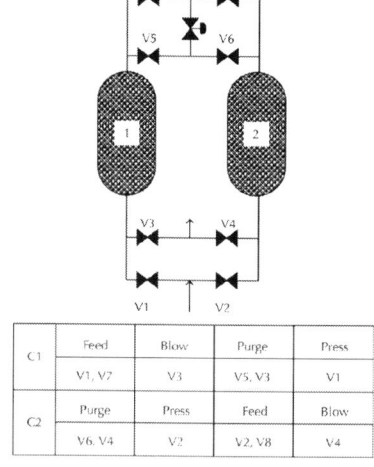

		Feed	Blow	Purge	Press
C1		V1, V7	V3	V5, V3	V1
		Purge	Press	Feed	Blow
C2		V6, V4	V2	V2, V8	V4

Figure 1: Schematic design of the first two-column pressure swing adsorption unit and valve sequencing for different steps in the cycle [1].

The four steps of the "Skarstrom cycle" are also shown in Figure 1: feed, blowdown (or evacuation), purge and pressurization. In this cycle, in the feed step, air is fed to the first column (C1) at a pressure higher than atmospheric. The adsorbent initially used (zeolite 5A) is selective to nitrogen, making the exiting stream (after valve V7) richer in oxygen. When the adsorbent packed in C1 is saturated and cannot adsorb more nitrogen, the feed is directed to the second column (C2). In order to release part of the nitrogen adsorbed in C1, the flow direction is reversed and the total pressure of the column is reduced by venting to atmosphere (opening valve V3). There are different terms to call this step, but blowdown is one of the most common and will be used here. In the blowdown step, nitrogen is desorbed from the adsorbent and released and at the end of this step, the gas phase inside the column is rich in nitrogen. To additionally remove nitrogen from the column, a purge step (or light gas recycle) is used. The purge consists of recycled part of the enriched air from the other column which is flowing by the pressure differential between the two columns. After the adsorbent is ready to load more nitrogen, the overall pressure of the system should be restored. That is done in the pressurization step using the feed stream. After all these steps were finished, a complete cycle was completed. It is important to notice that although the column operation is discontinuous, the feed stream is being employed so the process can be viewed as continuous. However, the exit is discontinuous and a tank is required to be coupled for a continuous discharge. Also, the operation in both columns should be synchronized to satisfy the continuous utilization of the feed and to provide purge gas to the other column.

The requirement of continuous feed processing, even being a discontinuous process, was recognized, since one of the first inventions of adsorption processes [105]. Furthermore, the valve arrangement for sequential opening - close and step definition was also very similar to designs presented for TSA processes [106]. However, the contribution of Skarstrom allowed a tremendous improvement in utilization of the adsorbents: while TSA cycles last for several hours, the PSA cycles are much shorter and thus using more adsorbent per unit time.

Another important aspect of a PSA process was mentioned in Skarstrom's application: heat effects and conservation. In the adsorption step, heat generated by adsorption may be important in which case the temperature of the column changes with time and also with position [4,

5, 55]. The consequence is a reduction in the adsorbent capacity. The "heat effects" may be very important in designing a PSA unit [107] and should be taken into account in the design: laboratory or small-scale experiments are either isothermal or close to isothermal and the heat capacity of the wall is important while large-scale processes behave adiabatically. In the desorption steps, the opposite is happening: energy is required for desorption resulting in a temperature decrease enhancing the potential capacity of the adsorbent and making desorption more difficult. This will happen in all PSA applications but in some cases, the amount of heat generated is not so important and the process can be considered isothermal. Every time there is a temperature swing associated to the PSA cycle, the performance is worse than what would be if the cycle is isothermal. However, since the thermal effects are present, it is good practice to conserve the "heat wave" inside the column: this heat will be used for a faster desorption.

MODIFICATIONS TO THE SKARSTROM CYCLE: NEW CYCLE STEPS

In the years after Skarstrom invention, there were several patent applications to improve the cycle. In a patent that was filled almost at the same time as Skarstrom, the regeneration under vacuum was introduced by Guerin de Montgareuil and Domine [73]. When vacuum is used for regeneration it is common to term the unit as vacuum pressure swing adsorption (VPSA). Although the utilization of vacuum may have an impact on the energetic requirements of the system, the efficiency of the unit may be greatly improved if the loading of the most adsorbed components changes dramatically at pressure lower than atmospheric. In the same invention, the authors have introduced the utilization of the pressurization step using part of the enriched gas. The utilization of a pressurization using part of the purified gas had impact in the purity of the produced gas [108]. Even when using the same pressure swing concept, the alternatives to develop the PSA technology are quite diverse, opening the "PSA engineering" possibilities.

The introduction of a pressure equalization step was developed at ESSO Research group [74, 109, 110]. Taking the two-column PSA scheme from Figure 1, after C1 ends the feed step (and is at high pressure), C2 ends the purge step (and is at low pressure). In that moment, V5 and V6 are simultaneously opened, short-circuiting the columns. This means that part of the gas that will normally get lost in the blowdown step is being used to pressurize the other column, loosing less purified gas. If the gas moving from one column to the other is not significantly adsorbed (e.g., hydrogen) the pressure achieved after the equalization step is the geometric average between these two values. The overall pressure can be lower if the gas transferred is fast adsorbed [111]. The result of the pressure equalization step is a direct improvement in the recovery of the light product [112, 113]. The introduction of a pressure equalization step in a 2-column PSA unit results in a significant change of the "continuity" of the process. When the two columns are in pressure equalization, there is no feed processing so at least one more column is required [110].

When several columns are employed, several pressure equalization steps can be done [114–116] and as a consequence, the overall recovery is increased [65, 117, 118]. This finding resulted in the design of multiple column (Polybed) PSA units [65].

Another possibility to remove part of the light component from the column before blowdown is depressurizing the bed co-currently to the feed direction. This step is very useful in hydrogen purification and is normally termed as "provide purge" step since it provides gas for purging other column [119].

Co-n-current depressurization was also used to remove the less adsorbed gas from the column in order to increase the content of most-adsorbed gas inside the column (aiming to its concentration) [32, 120–122].

An interesting concept of column depressurization is provided by the unique availability of "free vacuum" obtained in outer space [123]. In order to have a faster depressurization, it was proposed to open the column from both ends to release the gas faster. Parallel equalization using valves at different column lengths were also suggested [124]. Using a low-pressure feed as a purge was also suggested to increase the purity and recovery when compared with the Skarstrom cycle [125]. For the case of separation of a ternary mixture, feeding and one product withdrawal in intermediate positions of the column was also suggested, with a PSA design resembling the Petliuk scheme for distillation [126, 127].

In order to displace the light component to the product end, a recycle of the heavy component was suggested by Basmadjian and Pogorski [128]. This step was called "rinse." Although the rinse step aimed to provide a solution to concentration of low-per cent light compounds, it has been widely used for other purposes: concentration of the more adsorbed species [32, 120–122, 129–132].

In fact, the number of possible "steps" is not very large. However, using them in an efficient way has proved to be a difficult task. So far, the question raised by Professor Ruthven in 1992 was not yet completely answered [133] ("Is it possible to develop an algorithm for automatic generation of PSA cycles and tuning of the various steps?").

PERFORMANCE INDICATOR PARAMETERS OF A PSA PROCESS

So far, it has been shown that PSA processes have a tremendous flexibility in design (so large that sometimes is misleading). A completely different number of columns can be used and also a quite large number of cycles are possible. In order to provide a certain "common framework" to understand some aspects of PSA engineering, it is desirable to have some "performance indicators" (PI) which will be the ones that will define how well performs the PSA process. For the definition of such parameters, the PSA process depicted in Figure 2 can be considered. The image shows a PSA process with X columns (X can also be unity) accommodating a specific mass of adsorbent per column (w_{ads}) and with multiple connection lines to accommodate very different steps. The objective is to separate component i from N components and two cases may be found: either the purpose of the PSA is to purify the less adsorbed gas or alternatively, to concentrate the more adsorbed gas.

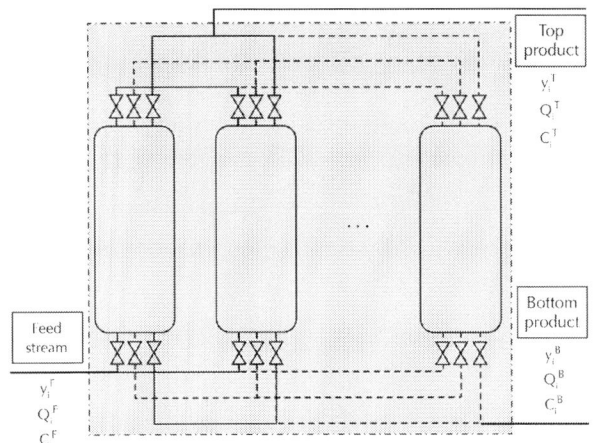

Figure 2: "Grey-box" generic example of a pressure swing adsorption (PSA) process. The inlet and exit streams are characterized by molar fraction (y_i) volumetric flowrate (Q_i, m³) and gas concentration (C_i, mol/m³).

The most common PI found in PSA processes are listed in Table 1 [134]. The two first PI (purity and recovery) are related to the separation efficiency of the PSA and normally establish the GO/NO GO condition in process design. If such specifications are satisfied, the "fingerprint" of the unit is evaluated by the productivity. Finally, the energetic considerations should be made. Since the process is so flexible, it is difficult to define an energetic PI other than saying that is the sum of all work used for compression and vacuum. Note that the recovery and productivity have an integral term that is mainly due to variations in flowrate in the exit streams.

Table 1: Performance indicator parameters for a PSA process

Less-adsorbed gas is the product	More-adsorbed gas is the product
$\text{Purity} = \dfrac{C_i^T}{\sum_{i=1}^{N} C_i^T} = y_i^T$	$\text{Purity} = \dfrac{C_i^B}{\sum_{i=1}^{N} C_i^B} = y_i^B$

Recovery = $\dfrac{\int_0^t C_i^T Q_i^T \, dt}{C_i^F Q_i^F}$	Recovery = $\dfrac{\int_0^t C_i^B Q_i^B \, dt}{C_i^F Q_i^F}$
Productivity = $\dfrac{\int_0^{t_{cycle}} C_i^T Q_i^T \, dt}{X w_{ads} t_{cycle}}$	Productivity = $\dfrac{\int_0^{t_{cycle}} C_i^B Q_i^B \, dt}{X w_{ads} t_{cycle}}$
Energy = sum of all compression and vacuum sources used	

Most works on PSA processes have shown that normally the purity and recovery present a trade-off for the design. In the case of recovering the less adsorbed gas, if more purge is used, more of the contaminants can be desorbed from the column and purity increases, but since more light gas is exiting from the "bottom end," light-gas recovery is smaller. A similar effect is observed for the utilization of the rinse step and purity and recovery of the more adsorbed gas.

However, other strategies are valid to improve process recovery without seriously affecting the purity. The case of Polybed PSA for H_2 purification is a good example [65]. The units built until 1975 were having 4 columns and the recovery of H_2 was around 60%. Nowadays, PSA unit with 12 columns are found [65] and up to 16 columns were patented [135] with H_2 recovery close to 90%. When the number of columns is increased, more pressure equalization steps can be performed and thus less hydrogen is lost with the contaminants, increasing its recovery.

The developments in the PSA process presented above were mainly motivated to improve the purity and the recovery of the target product(s). Nowadays, several new applications of PSA as an alternative technology are still in the stage of finding proper cycle configuration (step scheduling and times, number of columns, etc.). Other applications in more established markets are intending to improve either the unit size and/or the energetic consumption of the separation.

THE ROLE OF THE ADSORBENT IN PSA

The development of materials science in the last 60 years was quite intense. The result was the discovery of many porous materials, from all kind of zeolites and mesoporous materials [136–141] to the most diverse surfaces in activated carbons [142–145] and lately the high-surface area coordination polymers [146–151]. However, as strange as it may seem, only few materials are used in PSA units nowadays.

A review of adsorption properties of the different materials is out of the scope of this work, but good databases with adsorption properties of different gases on several adsorbents can be found [16, 152, 153]. What is important to mention is that a material to be used in PSA should be easily regenerated. It is frequent to find in literature adsorbents with a very high capacity, particularly at low pressures. Normally the isotherms of gases on such adsorbents are "rectangular": very steep at low pressures and quite flat after a certain pressure. Defining the "cyclic capacity" as the difference of loading between the high and low pressures of the PSA cycle, the only way to have an acceptable cyclic capacity is making blowdown at very high vacuum. The direct implication of using such conditions is that the power consumption increases rapidly. So, materials showing linear or slightly nonlinear isotherms are preferred in PSA design.

One frequent case is to have a multicomponent mixture of gases and that the number of compounds to be separated cannot be removed by a single adsorbent. The solution to this problem was found for the case of H_2 purification from methane steam reforming. In this application, H_2 is mixed with H_2O, CO_2, CO, unconverted CH_4, and possibly other gases like N_2. Activated carbon can be used to remove H_2O and CO_2 quite selectively but the loading for CO is rather limited for small partial pressures. It is thus common practice to use different layers of adsorbents to increase the loading of CO in the same column. This approach has also been applied in other separations [66, 70, 79, 154–160]. Consecutive layers of adsorbents can also be used to improve the productivity of kinetic adsorbents by adding a material that can be easily regenerated after the kinetic adsorbent [161, 162].

Other important aspect regarding the material properties for PSA applications is the diffusion of the different gases through its porous structure. There are different types of "resistances" to diffuse from the bulk gas phase to the adsorption site [4, 5]. They are: boundary layer around the adsorbent particle, and resistances in the macro-meso pores, mouth of the micropores, and micropores (or crystals).

In some applications however, these mass transfer "problems" have become part of the solution. In fact, if the diffusional resistance of one of the components of the mixture is very large, this gas will take so long to adsorb that can be separated from other gas that diffuses faster through the pores.

The "kinetic processes" were recognized soon [28]. In fact, materials like zeolites are called "molecular sieves" because of this effect [136]. Another example of kinetic materials is the carbon molecular sieves (CMS) [29–31, 33, 38, 163–167]. A CMS is prepared by contracting the pores of an activated carbon to limit the adsorption of some molecules. Its first utilization was for air separation to separate O_2 from N_2.

An extreme example of resistance to diffusion is the molecular exclusion like in the Isosiv process [5,97–99]. In the Isosiv process, n-paraffins are selectively adsorbed in zeolite 5A, while isoparaffins are kinetically excluded from the zeolite crystals.

Most recently, several inorganic materials have proved to be useful for kinetic separations [34, 36, 168–173]. A special kind of titanosilicates, ETS-4, cation exchanged with alkali-earth metals can be used for kinetic separations [35, 41, 174, 175]. In these materials, the pore size can be tuned with a very high accuracy by thermal treatment of the sample. Many studies have confirmed that CH_4 can be excluded from the structure while gases like H_2S, CO_2, and specially N_2 can be adsorbed [43, 176, 177].

ADVANCES IN PROCESS ENGINEERING

From all the main advances in process engineering, the most challenging one is the development of cyclic strategies that can improve the performance indicators of the PSA. Despite the performance of the material, the design of a PSA process requires several engineering decisions that should be taken sometimes with a very deep impact in terms of performance indicators. The main drawback of the engineering of a PSA process is that it is quite task consuming (and normally iterative).

With modern computers, the design of the PSA cycle can be carried out by modelling different scenarios. There are different degrees of complexity to define a PSA model, normally comprising several partial differential equations linked by the equation of state and the isotherm model to define the thermodynamic properties of the gas and adsorbed phases, respectively. Although the model can be solved by numerical methods [55, 113, 178–183], there are several commercial programs that can be already used for that purpose: ASPEN, COMSOL, gPROMS, PROSIM, and so forth [18, 184–187].

The simulation of a PSA process requires an initial step of defining a cycle structure (ordering the steps in a pre-defined sequence) and then estimate the performance indicators obtained. For the selected cycle, all the step times, blowdown pressure, and flowrates of rinse and purge steps should be determined [25, 188–192]. Alternatively, it has been suggested that a general "super-cycle" can be used to estimate the optimal duration of each of the steps [193].

In most cases, the definition of the cycle has to be done under certain constrains like combining it in a multiple column array. Other constraints can result from the availability of gas to the purge step, the continuous utilization of vacuum pump for blowdown, and so forth. The availability of gas to the purge step can also proceed from a depressurization step (provide purge) [119] or from a prestored amount in a tank [194]. A graphical procedure to schedule PSA cycles was suggested [195, 196]. It is also found in literature that in some cases, the best cycle does not match perfectly in a continuous array of columns and thus an "idle" step is used where the column is closed and no effective step for adsorption or desorption takes place. However, the existence of idle periods does result in smaller unit productivity of the PSA unit.

Recasting how the PSA productivity is calculated, we can see the interaction between the influence of process engineering and adsorbent development is mixed. If we have an adsorbent with a better cyclic capacity, we will be able to adsorb more gas per cycle and thus reduce the overall weight of adsorbent (or alternatively, increase the production of gas). On the other side, by better process engineering, we could improve the performance of the unit by balancing the amount of gas produced and possibly reducing the number of columns employed.

Furthermore, there is a third alternative: reduce the total cycle time. This alternative was suggested many years ago [197] and has started been implemented in the 80s [198]. When the total cycle time is smaller than 30 seconds, the process is normally called Rapid PSA (RPSA) [145, 179, 198–214].

A typical cycle time (t_{cycle}) of a normal PSA process is in the order of 10 minutes. In that time, the adsorbent is used to adsorb and desorb a certain amount of gas. Within each column of the PSA that amount adsorbed will be distributed in an initial zone where equilibrium has been achieved and a "mass transfer" zone close to the end of the column where the adsorbent is not completely saturated. The mass transfer zone is related to kinetic limitations to diffuse into the adsorbent and axial dispersion. Reducing the cycle time will result in more kinetic limitations and thus longer mass transfer zones. However, if reducing the cycle time in a factor of 10 results in a decrease of the amount adsorbed/desorbed in a factor of 2 (by kinetic limitations to adsorb), then the overall productivity of the PSA unit has still increased in a factor of 5. The result is that the PSA unit will be five times smaller!

There are several fields where RPSA can make a complete difference. A PSA for production of medicinal oxygen is a very suitable unit for utilization in hospitals. However, the concept of RPSA has opened the possibility of portable devices with quite small size that can be used for ambulatory patients with chronic lung diseases [78, 215]. Comparing the productivity of a PSA process to purify hydrogen, it can be noted that is quite lower than the productivity found in other PSA applications. In such a field, the utilization of RPSA concept can lead to significant reduction in size [201, 216].

The utilization of RPSA is limited by fluid dynamics. Using the Ultra-rapid piston driven PSA, the total cycle time was less than 5 seconds (its adsorption/desorption cycles resemble the expansion and compression of an internal combustion engine). Under such conditions, the mathematical models used to simulate normal PSA processes may not work [210, 217]: mass and energy transfer description using simplifications like LDF (linear driving force) are not applicable. There are also some particularities related to RPSA that could be overcome with the utilization of specialized devices.

In RPSA processes, the time required for pressurization of the bed can be a problem. It has been proven that by using a honeycomb monolith, it is possible to reduce the pressure drop of the PSA process [209] and thus reduce the overall pressurization time. Alternative to monolithic structures, laminated adsorbents have been suggested [218].

The other invention that is directly applicable to RPSA technology is the rotary valve [205, 207, 219]. Taking as example the PSA unit shown in Figure 1, it can be observed that the step changes in a normal PSA are accomplished by the simultaneous operation of a sometimes complex valve array. Using rotary multiport valves, it is possible to change the events taking place in all the columns at the exact same time. Using a normal valve array, a failure of one second in opening or closing one of the valves can have a significant impact in a RPSA cycle.

Another approach to PSA technology was carried out using radial columns [220–222]. Using radial columns, the length of adsorbent is normally small (resulting in decreased pressure drop) and the amount of gas to be treated at a reasonable gas velocity can be higher.

CONCLUDING REMARKS

The great flexibility of PSA is normally associated to process complexity and is still one of the major issues to introduce this technology in several fields of industry. On the other hand, the large flexibility of PSA processes still constitutes its main advantage and may be the reason why it has found applications in diverse fields.

PSA technology can be considered a mature technology in air separation, drying, and hydrogen purification, but there is plenty of work to do to establish this technique in other fields [223]. Many researchers around the world are currently working on CO_2 capture from flue gases. It has been potentially demonstrated that CO_2 can be captured using PSA [224–227] but more fundamental and long-term pilot plant studies are required to properly benchmark this technique against amines. Also, olefin-paraffin separation by adsorption was quite studied, but the energetic consumption of the separation by adsorption is still comparable to distillation [228]. Utilization of PSA for natural gas upgrading (CH_4-CO_2 separation basically) still also remains a challenge [229, 230]. PSA technology and even RPSA can be used to upgrade biogas, but the flowrate and pressure levels of natural gas require alternative solutions. Furthermore, new stringent legislation related to reducing the emission of greenhouse gases is changing the design of processes in energy and fuel industries. New processes intend to include or integrate the CO_2 capture, thus introducing specifications in the most adsorbed compound. A solution that is already in use and should be more explored is the dual PSA concept [231–235].

In all these emergent applications of PSA technology, faster and better solutions can happen by having a good interaction between materials science and process engineering.

REFERENCES

1. I. Langmuir, "The adsorption of gases on plane surfaces of glass, mica and platinum," The Journal of the American Chemical Society, vol. 40, no. 9, pp. 1361–1403, 1918.
2. M. Polanyi, "Section III.—theories of the adsorption of gases: a general survey and some additional remarks. Introductory paper to section III," Transactions of the Faraday Society, vol. 28, pp. 316–333, 1932.
3. A. L. Myers and J. M. Prausnitz, "Thermodynamics of mixed-gas adsorption," AIChE Journal, vol. 11, pp. 121–127, 1965.
4. D. M. Ruthven, Priciples of Adsorption and Adsorption Processes, John Wiley & Sons, New York, NY, USA, 1984.
5. R. T. Yang, Adsorbents. Fundamentals and Applications, John Wiley & Sons, New Jersey, NJ, USA, 2003.
6. P. C. Wankat, Large-Scale Adsorption and Chromatography, CRC Press, Boca Raton, Fla, USA, 1986.
7. A. E. Rodrigues, M. D. LeVan, and D. Tondeur, Adsorption, Science and Technology, Kluwer Academic Publishers, Boston, Mass, USA, 1989.
8. M. Suzuki, Adsorption Engineering, Chemical Engineering Monographs, Elsevier, Tokyo, Japan, 1990.
9. J. Kärger and D. M. Ruthven, Diffusion in Zeolites and Other Microporous Solids, John Wiley & Sons, London, UK, 1992.

10. C. Tien, Adsorption Calculations and Modeling, Butterworth-Heinemann, Boston, Mass, USA, 1994.
11. D. Basmadjian, The Little Adsorption Book: A Practical Guide for Engineers and Scientists, CRC Press, Boca Raton, Fla, USA, 1997.
12. J. L. Humphrey and G. E. Keller, Separation Process Technology, McGraw-Hill, New York, NY, USA, 1997.
13. D. D. Do, Adsorption Analysis: Equilibria and Kinetics, Imperial College Press, London, UK, 1998.
14. J. W. Thomas and B. D. Crittenden, Adsorption Technology and Design, Elsevier, Boston, Mass, USA, 1998.
15. O. Talu, "Needs, status, techniques and problems with binary gas adsorption experiments," Advances in Colloid and Interface Science, vol. 76-77, pp. 227–269, 1998.
16. F. Rouquerol, J. Rouquerol, and K. Song, Adsorption by Powders and Porous Solids, Academic Press, London, UK, 1999.
17. J. Keller and R. Staudt, Gas Adsorption Equilibria: Experimental Methods and Adsorption Isotherms, Springer, Boston, Mass, USA, 2005.
18. P. C. Wankat, Separation Process Engineering, Prentice Hall, London, UK, 2nd edition, 2007.
19. K. S. Knaebel and F. B. Hill, "Pressure swing adsorption: development of an equilibrium theory for gas separations," Chemical Engineering Science, vol. 40, no. 12, pp. 2351–2360, 1985.
20. M. D. LeVan, "Pressure swing adsorption: equilibrium theory for purification and enrichment," Industrial and Engineering Chemistry Research, vol. 34, no. 8, pp. 2655–2660, 1995.
21. G. Pigorini and M. D. LeVan, "Equilibrium theory for pressure swing adsorption. 2: purification and enrichment in layered beds," Industrial and Engineering Chemistry Research, vol. 36, no. 6, pp. 2296–2305, 1997.
22. G. Pigorini and M. D. LeVan, "Equilibrium theory for pressure swing adsorption. 3: separation and purification in two-component adsorption," Industrial and Engineering Chemistry Research, vol. 36, no. 6, pp. 2306–2319, 1997.
23. G. Pigorini and M. D. LeVan, "Equilibrium theory for pressure-swing adsorption. 4: optimizations for trace separation and

purification in two-component adsorption," Industrial and Engineering Chemistry Research, vol. 37, no. 6, pp. 2516–2528, 1998.

24. A. Serbezov and S. V. Sotirchos, "Semianalytical solution for multicomponent pressure swing adsorption," Chemical Engineering Science, vol. 53, no. 20, pp. 3521–3536, 1998. ·

25. A. Serbezov, "Effect of the process parameters on the lenght of the mass transfer zone during product withdrawal in pressure swing adsorption cycles," Chemical Engineering Science, vol. 56, no. 15, pp. 4673–4684, 2001. ·

26. A. D. Ebner and J. A. Ritter, "Equilibrium theory analysis of rectifying PSA for heavy component production," AIChE Journal, vol. 48, no. 8, pp. 1679–1691, 2002. ·

27. A. D. Ebner and J. A. Ritter, "Equilibrium theory analysis of dual reflux PSA for separation of a binary mixture," AIChE Journal, vol. 50, no. 10, pp. 2418–2429, 2004. ·

28. H. W. Habgood, "The kinetics of molecular sieve action: sorption of nitrogen-methane mixtures by Linde Molecular Sieve 4A," Canadian Journal of Chemistry, vol. 36, pp. 1384–1397, 1958.

29. K. Chihara, M. Suzuki, and K. Kawazoe, "Adsorption rate on molecular sieving carbon by chromatography," AIChE Journal, vol. 24, no. 2, pp. 237–246, 1978.

30. H. Jüntgen, K. Knoblauch, and K. Harder, "Carbon molecular sieves: production from coal and application in gas separation," Fuel, vol. 60, no. 9, pp. 817–822, 1981.

31. A. Kapoor and R. T. Yang, "Kinetic separation of methane-carbon dioxide mixture by adsorption on molecular sieve carbon," Chemical Engineering Science, vol. 44, no. 8, pp. 1723–1733, 1989.

32. R. Ramachandran, L. H. Dao, and B. Brooks, "Method of producing unsaturated hydrocarbons and separating the same from saturated hydrocarbons," U.S. patent 5, 365, 011, 1994.

33. A. I. Fatehi, K. F. Loughlin, and M. M. Hassan, "Separation of methane-nitrogen mixtures by pressure swing adsorption using a carbon molecular sieve," Gas Separation and Purification, vol. 9, no. 3, pp. 199–204, 1995. ·

34. M. W. Seery, "Bulk separation of carbon dioxide from methane using natural clinoptilolite," World Patent, WO 98/58726, 1998.
35. S. M. Kuznicki, V. A. Bell, I. Petrovic, and P. W. Blosser, "Separation of nitrogen from mixtures thereof with methane utilizing barium exchanged ETS-4," US patent no. 5, 989, 316, 1999.
36. J. Padin, S. U. Rege, R. T. Yang, and L. S. Cheng, "Molecular sieve sorbents for kinetic separation of propane/propylene," Chemical Engineering Science, vol. 55, no. 20, pp. 4525–4535, 2000. ·
37. M. Mitariten, "New technology improves nitrogen-removal economics," Oil and Gas Journal, vol. 99, no. 17, pp. 42–44, 2001.
38. A. Jayaraman, A. S. Chiao, J. Padin, R. T. Yang, and C. L. Munson, "Kinetic separation of methane/carbon dioxide by molecular sieve carbons," Separation Science and Technology, vol. 37, no. 11, pp. 2505–2528, 2002. ·
39. W. B. Dolan and M. J. Mitariten, "Heavy hydrocarbon recovery from pressure swing adsorption unit tail gas," 2003, US patent 6, 610, 124.
40. W. B. Dolan and M. J. Mitariten, "CO_2 rejection from natural gas," US patent 2003/0047071, 2003.
41. S. M. Kuznicki and V. A. Bell, "Olefin separation employing ETS molecular sieves," U.S. patent, 6, 517, 611, 2003.
42. M. B. Kim, Y. S. Bae, D. K. Choi, and C. H. Lee, "Kinetic separation of landfill gas by a two-bed pressure swing adsorption process packed with carbon molecular sieve: nonisothermal operation,"Industrial and Engineering Chemistry Research, vol. 45, no. 14, pp. 5050–5058, 2006. ·
43. S. Cavenati, C. A. Grande, F. V. S. Lopes, and A. E. Rodrigues, "Adsorption of small molecules on alkali-earth modified titanosilicates," Microporous and Mesoporous Materials, vol. 121, no. 1–3, pp. 114–120, 2009. ·
44. D. B. Shah, S. P. Perera, and B. D. Crittenden, "Adsorption dynamics in a monolithic adsorbent," inFundamentals of Adsorption, M. D. LeVan, Ed., Kluwer Academic Publishers, Boston, Mass, USA, 1996.
45. K. P. Gadkaree, "System and method for adsorbing contaminants and regenerating the adsorber," U.S. patent 5, 658, 372, 1997.

46. Y. Y. Li, S. P. Perera, and B. D. Crittenden, "Zeolite monoliths for air separation—part 2: oxygen enrichment, pressure drop and pressurization," Chemical Engineering Research and Design, vol. 76, no. 8, pp. 931–941, 1998. ·
47. R. Jain, A. I. LaCava, A. Maheshwary, J. R. Ambriano, D. R. Acharya, and F. R. Fitch, "Air separation using monolith adsorbent bed," U. S. patent, 6, 231, 644, 2001.
48. R. E. Critoph, "Multiple bed regenerative adsorption cycle using the monolithic carbon-ammonia pair," Applied Thermal Engineering, vol. 22, no. 6, pp. 667–677, 2002. ·
49. D. J. Kim, J. W. Kim, J. E. Yie, and H. Moon, "Temperature-programmed adsorption and characteristics of honeycomb hydrocarbon adsorbers," Industrial and Engineering Chemistry Research, vol. 41, no. 25, pp. 6589–6592, 2002.
50. T. Valdés-Solís, M. J. G. Linders, F. Kapteijn, G. Marbán, and A. B. Fuertes, "Adsorption and breakthrough performance of carbon-coated ceramic monoliths at low concentration of n-butane,"Chemical Engineering Science, vol. 59, no. 13, pp. 2791–2800, 2004. ·
51. A. B. Gorbach, M. Stegmaier, G. Eigenberger, J. Hammer, and H. G. Fritz, "Compact pressure swing adsorption processes-impact and potential of new-type adsorbent-polymer monoliths," Adsorption, vol. 11, no. 1, pp. 515–520, 2005. ·
52. C. A. Grande, S. Cavenati, P. Barcia, J. Hammer, H. G. Fritz, and A. E. Rodrigues, "Adsorption of propane and propylene in zeolite 4A honeycomb monolith," Chemical Engineering Science, vol. 61, no. 10, pp. 3053–3063, 2006. ·
53. I. Perdana, D. Creaser, I. Made Bendiyasa, Rochmadi, and B. Wikan Tyoso, "Modelling NOxadsorption in a thin NaZSM-5 film supported on a cordierite monolith," Chemical Engineering Science, vol. 62, no. 15, pp. 3882–3893, 2007. ·
54. F. Rezaei and P. Webley, "Optimum structured adsorbents for gas separation processes," Chemical Engineering Science, vol. 64, no. 24, pp. 5182–5191, 2009. ·
55. D. M. Ruthven, S. Farooq, and K. S. Knaebel, Pressure Swing Adsorption, VCH Publishers, New York, NY, USA, 1994.

56. D. Tondeur and P. C. Wankat, "Gas purification by PSA," Separation and Purification Methods, vol. 14, no. 2, pp. 157–212, 1985.
57. C. W. Skarstrom, "Method and apparatus for fractionating gas mixtures by adsorption," U.S. patent 2, 944, 627, 1960.
58. S. Sircar, "Separation of multicomponent gas mixtures," U.S. patent 4, 171, 206, 1979.
59. P. Cen and R. T. Yang, "Separation of a five-component gas mixture by pressure swing adsorption,"Separation Science and Technology, vol. 20, no. 9-10, pp. 725–747, 1985. ·
60. S. Sircar, "Fractionation of multicomponent gas mixtures by pressure swing adsorption," U.S. patent 4, 790, 858, 1988.
61. T. C. Golden, R. Kumar, and W. C. Kratz, "Hydrogen purification," US patent 4, 957, 514, 1990.
62. R. Kumar, "Adsorption process for recovering two high purity gas products from multicomponent gas mixtures," U.S. patent 4, 913, 709, 1990.
63. O. Bomard, J. Jutard, S. Moreau, and X. Vigor, "Method for purifying hydrogen based gas mixtures using a lithium-exchanged X zeolite," W.O. Patent 97/45363, 1997.
64. A. Malek and S. Farooq, "Hydrogen purification from refinery fuel gas by pressure swing adsorption,"AIChE Journal, vol. 44, no. 9, pp. 1985–1992, 1998.
65. J. Stöcker, M. Whysall, and G. Q. Miller, 30 Years of PSA Technology for Hydrogen Purification, UOP LLC, Des Plaines, Ill, USA, 1998.
66. C. H. Lee, J. Yang, and H. Ahn, "Effects of carbon-to-zeolite ratio on layered bed H_2 PSA for coke oven gas," AIChE Journal, vol. 45, no. 3, pp. 535–545, 1999.
67. S. Sircar, W. E. Waldron, M. B. Rao, and M. Anand, "Hydrogen production by hybrid SMR-PSA-SSF membrane system," Separation and Purification Technology, vol. 17, no. 1, pp. 11–20, 1999. ·
68. J. H. Park, J. N. Kim, and S. H. Cho, "Performance analysis of four-bed H_2 PSA process using layered beds," AIChE Journal, vol. 46, no. 4, pp. 790–802, 2000.

69. S. Sircar and T. C. Golden, "Purification of hydrogen by pressure swing adsorption," Separation Science and Technology, vol. 35, no. 5, pp. 667–687, 2000. · ·
70. J. G. Jee, M. B. Kim, and C. H. Lee, "Adsorption characteristics of hydrogen mixtures in a layered bed: binary, ternary, and five-component mixtures," Industrial and Engineering Chemistry Research, vol. 40, no. 3, pp. 868–878, 2001.
71. M. S. A. Baksh, M. W. Ackley, and F. Notaro, "Process and apparatus for hydrogen purification," W.O. Patent 2004/058630, 2004.
72. R. L. Bec, "Method for purifying hydrogen-based gas mixtures using calcium X-zeolite," U.S. patent 6, 849, 106, 2005.
73. P. Guerin de Montgareuil and D. Domine, "Process for separating a binary gaseous mixture by adsorption," US patent 3, 155, 468, 1964.
74. N. H. Berlin, "Method for providing an oxygen-enriched environment," U.S. Patent 3, 280, 536, 1966.
75. H. Jüntgen, K. Knoblauch, J. Reichenberger, and F. Tarnow, "Process for the recovery of nitrogen-rich gases from gases containing at least oxygen as other component," U.S. patent 4, 264, 339, 1981.
76. D. M. Ruthven, N. S. Raghavan, and M. M. Hassan, "Adsorption and diffusion of nitrogen and oxygen in a carbon molecular sieve," Chemical Engineering Science, vol. 41, no. 5, pp. 1325–1332, 1986.
77. C. G. Coe, J. F. Kirner, R. Pierantozzi, and T. R. White, "Nitrogen adsorption with Ca and or Sr exchanged lithium X-zeolites," U.S. patent 5, 152, 813, 1992.
78. J. G. Jee, J. S. Lee, and C. H. Lee, "Air separation by a small-scale two-bed medical O_2 pressure swing adsorption," Industrial and Engineering Chemistry Research, vol. 40, no. 16, pp. 3647–3658, 2001.
79. Y. Lü, S. J. Doong, and M. Bülow, "Pressure-swing adsorption using layered adsorbent beds with different adsorption properties: II-experimental investigation," Adsorption, vol. 10, no. 4, pp. 267–275, 2005. ·

80. J. C. Santos, A. F. Portugal, F. D. Magalhães, and A. Mendes, "Optimization of medical PSA units for oxygen production," Industrial and Engineering Chemistry Research, vol. 45, no. 3, pp. 1085–1096, 2006. ·
81. K. G. Teague and T. F. Edgar, "Predictive dynamic model of a small pressure swing adsorption air separation unit," Industrial and Engineering Chemistry Research, vol. 38, no. 10, pp. 3761–3775, 1999.
82. K. T. Chue, J. N. Kim, Y. J. Yoo, S. H. Cho, and R. T. Yang, "Comparison of activated carbon and zeolite 13X for CO_2 recovery from flue gas by pressure swing adsorption," Industrial and Engineering Chemistry Research, vol. 34, no. 2, pp. 591–598, 1995.
83. J. H. Park, H. T. Beum, J. N. Kim, and S. H. Cho, "Numerical analysis on the power consumption of the PSA process for recovering CO_2 from flue gas," Industrial and Engineering Chemistry Research, vol. 41, no. 16, pp. 4122–4131, 2002.
84. A. L. Chaffee, G. P. Knowles, Z. Liang, J. Zhang, P. Xiao, and P. A. Webley, "CO_2 capture by adsorption: materials and process development," International Journal of Greenhouse Gas Control, vol. 1, no. 1, pp. 11–18, 2007. ·
85. J. S. D׳amico, H. E. Reinhold III, and K. S. Knaebel, "Helium recovery," U.S. patent, 5, 542, 966, 1996.
86. N. K. Das, H. Chaudhuri, R. K. Bhandari, D. Ghose, P. Sen, and B. Sinha, "Purification of helium from natural gas by pressure swing adsorption," Current Science, vol. 95, no. 12, pp. 1684–1687, 2008.
87. A. P. G. Taveira and A. M. M. Mendes, "Xenon external recycling unit for recovery, purification and reuse of xenon in anesthesia circuits," U.S. patent 7, 442, 236, 2008.
88. S. Sircar and W. R. Kock, "Adsorptive separation of methane and carbon dioxide gas mixtures," European patent EP, 0193716, 1986.
89. S. Sircar, "High efficiency separation of methane and carbon dioxide mixtures by adsorption: adsorption and ion exchange," AIChE Symposium Series, vol. 84, pp. 70–72, 1988.

90. S. Sircar, R. Kumar, W. R. Koch, and J. Vansloun, "Recovery of methane from landfill gas," United States patent 4, 770, 676, 1988.

91. M. M. Davis, R. L. J. Gray, and K. Patei, "Process for the purification of natural gas," US patent, 5, 174, 796, 1992.

92. M. Mitariten, "Economic N_2 removal," Hydrocarbon Engineering, vol. 9, no. 7, pp. 53–57, 2004.

93. I. A. A. C. Esteves and J. P. B. Mota, "Simulation of a new hybrid membrane/pressure swing adsorption process for gas separation," Desalination, vol. 148, no. 17–3, pp. 275–280, 2002. ·

94. K. S. Knaebel and H. E. Reinhold, "Landfill gas: from rubbish to resource," Adsorption, vol. 9, no. 1, pp. 87–94, 2003. ·

95. C. A. Grande and A. E. Rodrigues, "Biogas to fuel by vacuum pressure swing adsorption I. Behavior of equilibrium and kinetic-based adsorbents," Industrial and Engineering Chemistry Research, vol. 46, no. 13, pp. 4595–4605, 2007. ·

96. C. A. Grande and R. Blom, "Utilization of Dual-PSA technology for natural gas upgrading and integrated CO_2 capture," Energy Procedia, vol. 26, pp. 2–14, 2012.

97. "Isosiv process operates commercially," Chemical & Engineering News, vol. 40, pp. 59–63, 1962.

98. T. C. Holcombe, "N-paraffin—isoparaffin separation process," U.S. patent, 4, 176, 053, 1979.

99. J. A. C. Silva, Separation of n/iso—paraffins by adsorption process [Ph.D. dissertation], University of Porto, Porto, Portugal, 1998.

100. A. Mersmann, B. Fill, R. Hartmann, and S. Maurer, "The potential of energy saving by gas phase adsorption processes," Chemical Engineering & Technology, vol. 23, pp. 937–944, 2000.

101. S. Sircar, "Pressure swing adsorption," Industrial and Engineering Chemistry Research, vol. 41, no. 6, pp. 1389–1392, 2002.

102. C. Voss, "Applications of pressure swing adsorption technology," Adsorption, vol. 11, no. 1, pp. 527–529, 2005. ·

103. K. Knaebel, "Adsorbent selection," 2004, http://www.adsorption.com/publications/ AdsorbentSel1B.pdf.

104. N. Sundaram and R. T. Yang, "On the pseudomultiplicity of pressure swing adsorption periodic states," Industrial and

Engineering Chemistry Research, vol. 37, no. 1, pp. 154–158, 1998.
105. W. A. Patrick, B. F. Lovelace, and E. B. Miller, "Method and apparatus for separating vapors and gases," U.S. patent 1, 335, 348, 1920.
106. A. B. Ray, "Process of recovering absorbable constituents from gas streams," U.S. patent 1, 548, 280, 1925.
107. R. T. Yang and P. L. Cen, "Improved pressure swing adsorption processes for gas separation: by heat exchange between adsorbers and by high-heat-capacity inert additives," Industrial & Engineering Chemistry Process Design and Development, vol. 25, no. 1, pp. 54–59, 1986. ·
108. H. Ahn, C. H. A. Lee, B. Seo, J. Yang, and K. Baek, "Backfill cycle of a layered bed H_2 PSA process,"Adsorption, vol. 5, no. 4, pp. 419–433, 1999. ·
109. W. D. Marsh, F. S. Pramuk, R. C. Hoke, and C. W. Skarstrom, "Pressure equalization depressurising in heatless adsorption," U.S. patent no. 3, 142, 547, 1964.
110. T. M. Stark, "Gas separation by adsorption process," U.S. patent 3, 252, 268, 1966.
111. M. P. S. Santos, C. A. Grande, and A. E. Rodrigues, "Pressure swing adsorption for biogas upgrading: effect of recycling streams in pressure swing adsorption design," Industrial and Engineering Chemistry Research, vol. 50, no. 2, pp. 974–985, 2011. ·
112. K. Warmuzinski, "Effect of pressure equalization on power requirements in PSA systems," Chemical Engineering Science, vol. 57, no. 8, pp. 1475–1478, 2002. · ·
113. J. A. Delgado and A. E. Rodrigues, "Analysis of the boundary conditions for the simulation of the pressure equalization step in PSA cycles," Chemical Engineering Science, vol. 63, no. 18, pp. 4452–4463, 2008. ·
114. J. L. Wagner, "Selective adsorption process," U.S. patent no. 3, 430, 418, 1969.
115. J. Xu, D. L. Rarig, T. A. Cook, K. K. Hsu, M. Schoonover, and R. Agrawal, "Pressure swing adsorption process with reduced pressure equalization time," US patent 6, 565, 628, 2003.

116. J. Xu and E. L. Weist Jr., "Six bed pressure swing adsorption process with four steps of pressure equalization," US patent 6, 454, 838, 2002.

117. N. Casas, J. Schell, and M. Mazzotti, "Pre-combustion CO_2 capture by PSA for IGCC plants," inProceedings of the 10th International Conference on Fundamentals of Adsorption (FOA '10), Awaji, Japan, May 2010.

118. F. V. S. Lopes, C. A. Grande, and A. E. Rodrigues, "Activated carbon for hydrogen purification by pressure swing adsorption: multicomponent breakthrough curves and PSA performance," Chemical Engineering Science, vol. 66, no. 3, pp. 303–317, 2011. · ·

119. A. Fuderer, "Pressure swing adsorption with intermediate product recovery," US patent 4, 512, 780, 1985.

120. R. Ramachandran and L. H. Dao, "Process for recovering alkenes from cracked hydrocarbon streams," US patent 5, 744, 687, 1998.

121. F. A. Da Silva and A. E. Rodrigues, "Propylene/propane separation by vacuum swing adsorption using 13X zeolite," AIChE Journal, vol. 47, no. 2, pp. 341–357, 2001. ·

122. C. A. Grande and A. E. Rodrigues, "Propane/propylene separation by pressure swing adsorption using zeolite 4A," Industrial and Engineering Chemistry Research, vol. 44, no. 23, pp. 8815–8829, 2005. ·

123. A. D. Ebner, J. A. Ritter, M. D. LeVan, and J. C. Knox, "Unique regeneration steps for the sorbent-based atmosphere revitalization system designed for CO_2 and H_2O removal from spacecraft cabins,"SAE International Journal of Aerospace, vol. 4, no. 1, pp. 488–493, 2011. ·

124. M. Yoshida, J. A. Ritter, A. Kodama, M. Goto, and T. Hirose, "Simulation of an enriching reflux PSA process with parallel equalization for concentrating a trace component in air," Industrial and Engineering Chemistry Research, vol. 45, no. 18, pp. 6243–6250, 2006. ·

125. K. P. Kostroski and P. C. Wankat, "High recovery cycles for gas separations by pressure-swing adsorption," Industrial and Engineering Chemistry Research, vol. 45, no. 24, pp. 8117–8133, 2006. ·

126. F. Dong, H. Lou, A. Kodama, M. Goto, and T. Hirose, "A new concept in the design of pressure-swing adsorption processes for multicomponent gas mixtures," Industrial and Engineering Chemistry Research, vol. 38, no. 1, pp. 233–239, 1999.

127. F. Dong, H. Lou, A. Kodama, M. Goto, and T. Hirose, "The Petlyuk PSA process for the separation of ternary gas mixtures: exemplification by separating a mixture of CO_2-CH_4-N_2," Separation and Purification Technology, vol. 16, no. 2, pp. 159–166, 1999.

128. D. Basmadjian and A. L. Pogorski, "Process for the separation of gases by adsorption," US patent 3, 279, 153, 1966.

129. T. Tamura, "Absorption process for gas separation," US patent 3, 797, 201, 1974.

130. D. Diagne, M. Goto, and T. Hirose, "Parametric studies on CO_2 separation and recovery by a dual reflux PSA process consisting of both rectifying and stripping sections," Industrial and Engineering Chemistry Research, vol. 34, no. 9, pp. 3083–3089, 1995.

131. B. K. Na, H. Lee, K. K. Koo, and H. K. Song, "Effect of rinse and recycle methods on the pressure swing adsorption process to recover CO_2 from power plant flue gas using activated carbon,"Industrial and Engineering Chemistry Research, vol. 41, no. 22, pp. 5498–5503, 2002.

132. S. P. Reynolds, A. D. Ebner, and J. A. Ritter, "Stripping PSA cycles for CO_2 recovery from flue gas at high temperature using a hydrotalcite-like adsorbent," Industrial and Engineering Chemistry Research, vol. 45, no. 12, pp. 4278–4294, 2006.

133. D. M. Ruthven, "PSA discussion," Studies in Surface Science and Catalysis, vol. 80, pp. 788–793, 1993.

134. R. Rota and P. C. Wankat, "Intensification of pressure swing adsorption processes," AIChE Journal, vol. 36, no. 9, pp. 1299–1312, 1990.

135. M. Whysall and L. J. M. Wagemans, "Very large-scale pressure swing adsorption processes," US patent 6, 210, 466, 2001.

136. W. D. Breck, Zeolite Molecular Sieves, John Wiley & Sons, New York, NY, USA, 1974.

137. J. E. Martin, M. T. Anderson, J. Odinek, and P. Newcomer, "Synthesis of periodic mesoporous silica thin films," Langmuir, vol. 13, no. 15, pp. 4133–4141, 1997. ·

138. D. Zhao, J. Feng, Q. Huo et al., "Triblock copolymer syntheses of mesoporous silica with periodic 50 to 300 angstrom pores," Science, vol. 279, no. 5350, pp. 548–552, 1998. ·

139. J. Rocha and M. W. Anderson, "Microporous titanosilicates and other novel mixed octahedral-tetrahedral framework oxides," European Journal of Inorganic Chemistry, no. 5, pp. 801–818, 2000.

140. S. Reyes, V. V. Krishnan, G. J. De Martin, J. H. Sinfelt, K. G. Strohmaier, and J. G. Santiesteban, "Separation of propylene from hydrocarbon mixtures," International patent, WO 03/080548 A1, 2003.

141. M. E. Rivera-Ramos, G. J. Ruiz-Mercado, and A. J. Hernández-Maldonado, "Separation of CO_2 from light gas mixtures using ion-exchanged silicoaluminophosphate nanoporous sorbents," Industrial and Engineering Chemistry Research, vol. 47, no. 15, pp. 5602–5610, 2008. ·

142. F. Rodríguez-Reinoso, M. Molina-Sabio, and M. T. González, "The use of steam and CO_2 as activating agents in the preparation of activated carbons," Carbon, vol. 33, no. 1, pp. 15–23, 1995.

143. Z. Liu, L. Ling, W. Qiao, and L. Liu, "Preparation of pitch-based spherical activated carbon with developed mesopore by the aid of ferrocene," Carbon, vol. 37, no. 4, pp. 663–667, 1999. ·

144. S. Jun, Sang Hoon Joo, R. Ryoo et al., "Synthesis of new, nanoporous carbon with hexagonally ordered mesostructure," Journal of the American Chemical Society, vol. 122, no. 43, pp. 10712–10713, 2000. ·

145. T. C. Golden, C. M. A. Golden, and D. P. Zwilling, "Self-supported structured adsorbent for gas separation," US patent 6, 565, 627, 2003.

146. J. L. C. Rowsell and O. M. Yaghi, "Metal-organic frameworks: a new class of porous materials," Microporous and Mesoporous Materials, vol. 73, no. 1-2, pp. 3–14, 2004. ·

147. P. D. C. Dietzel, Y. Morita, R. Blom, and H. Fjellvåg, "An in situ high-temperature single-crystal investigation of a dehydrated

metal-organic framework compound and field-induced magnetization of one-dimensional metal-oxygen chains," Angewandte Chemie—International Edition, vol. 44, no. 39, pp. 6354–6358, 2005. ·

148. U. Mueller, M. Schubert, F. Teich, H. Puetter, K. Schierle-Arndt, and J. Pastré, "Metal-organic frameworks—prospective industrial applications," Journal of Materials Chemistry, vol. 16, no. 7, pp. 626–636, 2006. ·

149. S. Ma, D. Sun, X. S. Wang, and H. C. Zhou, "A mesh-adjustable molecular sieve for general use in gas separation," Angewandte Chemie—International Edition, vol. 46, no. 14, pp. 2458–2462, 2007. ·

150. S. Cavenati, C. A. Grande, A. E. Rodrigues, C. Kiener, and U. Müller, "Metal organic framework adsorbent for biogas upgrading," Industrial and Engineering Chemistry Research, vol. 47, no. 16, pp. 6333–6335, 2008. ·

151. J. H. Cavka, S. Jakobsen, U. Olsbye et al., "A new zirconium inorganic building brick forming metal organic frameworks with exceptional stability," Journal of the American Chemical Society, vol. 130, no. 42, pp. 13850–13851, 2008. ·

152. D. P. Valenzuela and A. L. Myers, Adsorption Equilibrium Data Handbook, Prentice Hall, New Jersey, NJ, USA, 1989.

153. R. T. Yang, Adsorbents. Fundamentals and Applications, John Wiley & Sons, New Jersey, NJ, USA, 2003.

154. G. Klein and T. Vermeulen, "Cyclic performance of layered beds for binary ion exchange," AIChE Symposium Series, vol. 71, no. 15, pp. 69–76, 1975.

155. M. Chlendi and D. Tondeur, "Dynamic behaviour of layered columns in pressure swing adsorption," Gas Separation and Purification, vol. 9, no. 4, pp. 231–242, 1995. ·

156. C. F. Watson, R. D. Whitley, and M. L. Meyer, "Multiple zeolite adsorbent layers in oxygen separation," US patent 5, 529, 610, 1996.

157. J. H. Park, J. N. Kim, S. H. Cho, J. D. Kim, and R. T. Yang, "Adsorber dynamics and optimal design of layered beds for multicomponent gas adsorption," Chemical Engineering Science, vol. 53, no. 23, pp. 3951–3963, 1998.

158. J. Yang and C. H. Lee, "Adsorption dynamics of a layered bed PSA for H_2 recovery from coke oven gas," AIChE Journal, vol. 44, no. 6, pp. 1325–1334, 1998.
159. Y. Lü, S. J. Doong, and M. Bülow, "Pressure-swing adsorption using layered adsorbent beds with different adsorption properties: I—results of process simulation," Adsorption, vol. 9, no. 4, pp. 337–347, 2003. ·
160. S. Cavenati, C. A. Grande, and A. E. Rodrigues, "Separation of $CH_4/CO_2/N_2$ mixtures by layered pressure swing adsorption for upgrade of natural gas," Chemical Engineering Science, vol. 61, no. 12, pp. 3893–3906, 2006. ·
161. C. A. Grande and A. E. Rodrigues, "Layered vacuum pressure-swing adsorption for biogas upgrading," Industrial and Engineering Chemistry Research, vol. 46, no. 23, pp. 7844–7848, 2007. ·
162. C. A. Grande, S. Cavenati, and A. E. Rodrigues, "Separation column and pressure swing adsorption process for gas purification," World Patent Application, 2008/072215, 2008.
163. S. N. Vyas, S. R. Patwardhan, S. Vijayalakshmi, and K. S. Ganesh, "Adsorption of gases on carbon molecular sieves," Journal of Colloid And Interface Science, vol. 168, no. 2, pp. 275–280, 1994. ·
164. R. Srinivasan, S. R. Auvil, and J. M. Schork, "Mass transfer in carbon molecular sieves-an interpretation of Langmuir kinetics," The Chemical Engineering Journal, vol. 57, no. 2, pp. 137–144, 1995.
165. S. Farooq, H. Qinglin, and I. A. Karimi, "Identification of transport mechanism in adsorbent micropores from column dynamics," Industrial and Engineering Chemistry Research, vol. 41, no. 5, pp. 1098–1106, 2002.
166. H. Qinglin, S. M. Sundaram, and S. Farooq, "Revisiting transport of gases in the micropores of carbon molecular sieves," Langmuir, vol. 19, no. 2, pp. 393–405, 2003. ·
167. D. Shen, M. Bülow, and N. O. Lemcoff, "Mechanisms of molecular mobility of oxygen and nitrogen in carbon molecular sieves," Adsorption, vol. 9, no. 4, pp. 295–302, 2003. ·
168. M. W. Ackley and R. T. Yang, "Diffusion in ion-exchanged clinoptilolites," AIChE Journal, vol. 37, no. 11, pp. 1645–1656, 1991.

169. L. Predescu, F. H. Tezel, and S. Chopra, "Adsorption of nitrogen, methane, carbon monoxide, and their binary mixtures on aluminophosphate molecular sieves," Adsorption, vol. 3, no. 1, pp. 7–25, 1996.
170. W. Zhu, F. Kapteijn, J. A. Moulijn, M. C. Den Exter, and J. C. Jansen, "Shape selectivity in adsorption on the all-silica DD3R," Langmuir, vol. 16, no. 7, pp. 3322–3329, 2000. ·
171. D. H. Olson, "Light hydrocarbon separation using 8-member ring zeolites," US patent 6, 488, 741, 2002.
172. A. Jayaraman, A. J. Hernandez-Maldonado, R. T. Yang, D. Chinn, C. L. Munson, and D. H. Mohr, "Clinoptilolites for nitrogen/methane separation," Chemical Engineering Science, vol. 59, no. 12, pp. 2407–2417, 2004. ·
173. J. Gascón, W. Blom, A. van Miltenburg, A. Ferreira, R. Berger, and F. Kapteijn, "Accelerated synthesis of all-silica DD3R and its performance in the separation of propylene/propane mixtures,"Microporous and Mesoporous Materials, vol. 115, no. 3, pp. 585–593, 2008. ·
174. S. M. Kuznicki, "Preparation of small-pored crystalline titanium molecular sieve zeolites," US patent 4, 938, 939, 1991.
175. S. M. Kuznicki, V. A. Bell, S. Nair et al., "A titanosilicate molecular sieve with adjustable pores for size-selective adsorption of molecules," Nature, vol. 412, no. 6848, pp. 720–724, 2001. ·
176. J. H. Wills, M. Shemaria, and M. J. Mitariten, "Production of pipeline-quality natural gas with the molecular gate CO_2 removal process," SPE Production and Facilities, vol. 19, no. 1, pp. 4–8, 2004.
177. R. P. Marathe, K. Mantri, M. P. Srinivasan, and S. Farooq, "Effect of ion exchange and dehydration temperature on the adsorption and diffusion of gases in ETS-4," Industrial and Engineering Chemistry Research, vol. 43, no. 17, pp. 5281–5290, 2004.
178. O. J. Smith IV and A. W. Westerberg, "Mixed-integer programming for pressure swing adsorption cycle scheduling," Chemical Engineering Science, vol. 45, no. 9, pp. 2833–2842, 1990.
179. S. Farooq, C. Thaeron, and D. M. Ruthven, "Numerical simulation of a parallel-passage piston-driven PSA unit," Separation and Purification Technology, vol. 13, no. 3, pp. 181–193, 1998. ·

180. F. A. Da Silva, J. A. Silva, and A. E. Rodrigues, "General package for the simulation of cyclic adsorption processes," Adsorption, vol. 5, no. 3, pp. 229–244, 1999. ·
181. L. T. Biegler, L. Jiang, and V. G. Fox, "Recent advances in simulation and optimal design of pressure swing adsorption systems," Separation and Purification Reviews, vol. 33, no. 1, pp. 1–39, 2004. ·
182. P. A. Webley and J. He, "Fast solution-adaptive finite volume method for PSA/VSA cycle simulation; 1 single step simulation," Computers and Chemical Engineering, vol. 23, no. 11-12, pp. 1701–1712, 2000. ·
183. L. Jiang, V. G. Fox, and L. T. Biegler, "Simulation and optimal design of multiple-bed pressure swing adsorption systems," AIChE Journal, vol. 50, no. 11, pp. 2904–2917, 2004. ·
184. S. Nilchan and C. C. Pantelides, "On the optimisation of periodic adsorption processes," Adsorption, vol. 4, no. 2, pp. 113–147, 1998.
185. D. Nikolic, A. Giovanoglou, M. C. Georgiadis, and E. S. Kikkinides, "Generic modeling framework for gas separations using multibed pressure swing adsorption processes," Industrial and Engineering Chemistry Research, vol. 47, no. 9, pp. 3156–3169, 2008. · ·
186. D. Nikolic, E. S. Kikkinides, and M. C. Georgiadis, "Optimization of multibed pressure swing adsorption processes," Industrial and Engineering Chemistry Research, vol. 48, no. 11, pp. 5388–5398, 2009. ·
187. V. Rama Rao, S. Farooq, and W. B. Krantz, "Design of a two-step pulsed pressure-swing adsorption-based oxygen concentrator," AIChE Journal, vol. 56, no. 2, pp. 354–370, 2010. ·
188. N. Sundaram and P. C. Wankat, "Pressure drop effects in the pressurization and blowdown steps of pressure swing adsorption," Chemical Engineering Science, vol. 43, no. 1, pp. 123–129, 1988.
189. R. Kumar, "Adsorption column blowdown: adiabatic equilibrium model for bulk binary gas mixtures," Industrial and Engineering Chemistry Research, vol. 28, no. 11, pp. 1677–1683, 1989.
190. Z. P. Lu, J. M. Loureiro, A. E. Rodrigues, and M. D. LeVan, "Pressurization and blowdown of adsorption beds-II. Effect of the momentum and equilibrium relations on isothermal

operation,"Chemical Engineering Science, vol. 48, no. 9, pp. 1699–1707, 1993.
191. W. E. Waldron and S. Sircar, "Parametric study of a pressure swing adsorption process," Adsorption, vol. 6, no. 2, pp. 179–188, 2000. ·
192. D. Ko, R. Siriwardane, and L. T. Biegler, "Optimization of a pressure-swing adsorption process using zeolite 13X for CO_2 sequestration," Industrial and Engineering Chemistry Research, vol. 42, no. 2, pp. 339–348, 2003.
193. A. Agarwal, L. T. Biegler, and S. E. Zitney, "A superstructure-based optimal synthesis of PSA cycles for post-combustion CO_2 capteffectively captureure," AIChE Journal, vol. 56, no. 7, pp. 1813–1828, 2010. ·
194. K. Ramachandran, S. L. Lerner, and D. L. MacLean, "PSA multicomponent separation utilizing tank equalization," US patent 4, 816, 039, 1989.
195. A. Mehrotra, A. D. Ebner, and J. A. Ritter, "Arithmetic approach for complex PSA cycle scheduling,"Adsorption, vol. 16, no. 3, pp. 113–126, 2010. ·
196. A. Mehrotra, A. D. Ebner, and J. A. Ritter, "Simplified graphical approach for complex PSA cycle scheduling," Adsorption, vol. 17, no. 2, pp. 337–345, 2011. · ·
197. P. H. Turnock and R. H. Kadlec, "Separation of nitrogen and methane via periodic adsorption,"AIChE Journal, vol. 17, pp. 335–342, 1971.
198. R. L. Jones, I. I. Keller, I. I. G. E, and R. C. Wells, "Rapid pressure swing adsorption process with high enrichment factor," US patent 4, 194, 892, 1980.
199. D. E. Earls and G. N. Long, "Multiple bed rapid pressure swing adsorption for oxygen," US patent 4, 194, 891, 1980.
200. T. J. Dangieri and R. T. Cassidy, "Enhanced performance in rapid pressure swing adsorption processing," W.O. patent 86/002015, 1986.
201. S. Sircar, "Gas separation by rapid pressure swing adsorption," US patent 5, 071, 449, 1991.
202. S. Sircar and B. F. Hanley, "Production of oxygen enriched air by rapid pressure swing adsorption,"Adsorption, vol. 1, no. 4, pp. 313–320, 1995. ·

203. B. H. L. Betlem, R. W. M. Gotink, and H. Bosch, "Optimal operation of rapid pressure swing adsorption with slop recycling," Computers and Chemical Engineering, vol. 22, supplement 1, pp. S633–S636, 1998.
204. S. Kulish and R. P. Swank, "Rapid cycle pressure swing adsorption oxygen concentration method and apparatus," US patent 5, 827, 358, 1998.
205. B. G. Keefer, "High frequency pressure swing adsorption," U.S. Patent 6, 176, 897, 2001.
206. R. Arvind, S. Farooq, and D. M. Ruthven, "Analysis of a piston PSA process for air separation,"Chemical Engineering Science, vol. 57, no. 3, pp. 419–433, 2002. ·
207. D. J. Connor, D. G. Doman, L. Jeziorowski et al., "Rotary pressure swing adsorption apparatus," US patent 6, 406, 523, 2002.
208. T. C. Golden, E. L. Weist Jr., and P. A. Novosat, "Adsorbents for rapid cycle pressure swing adsorption processes," US patent 7, 404, 846, 2008.
209. E. M. Kopaygorodsky, V. V. Guliants, and W. B. Krantz, "Predictive dynamic model of single-stage ultra-rapid pressure swing adsorption," AIChE Journal, vol. 50, no. 5, pp. 953–962, 2004. ·
210. R. S. Todd and P. A. Webley, "Mass-transfer models for rapid pressure swing adsorption simulation,"AIChE Journal, vol. 52, no. 9, pp. 3126–3145, 2006. ·
211. S. Alizadeh-Khiavi, J. A. Sawada, A. C. Gibbs, and J. Alvaji, "Rapid cycle syngas pressure swing adsorption system," US patent application 2007/0125228, 2007.
212. T. C. Golden and E. L. Weist, "Activated carbon as sole absorbent in rapid cycle hydrogen PSA," US patent 6, 660, 064, 2003.
213. M. J. LaBuda, T. C. Golden, R. D. Whitley, and C. E. Steigerwalt, "Performance stability in rapid cycle pressure swing adsorption systems," European Patent Application, vol. 1, pp. 917–994, 2008.
214. N. Sundaram, B. K. Kaul, E. W. Corcoran, C. Y. Sabottke, and R. L. Eckes, "Integration of rapid cycle pressure swing adsorption with refinery process units (hydroprocessing, hydrocracking, etc.)," US patent 7, 591, 879, 2009.

215. C. Siew-Wah, S. Sircar, and M. V. Kothare, "Miniature oxygen concentrators and methods," US patent 8, 226, 745, 2012.

216. F. V. S. Lopes, C. A. Grande, and A. E. Rodrigues, "Fast-cycling VPSA for hydrogen purification," Fuel, vol. 93, pp. 510–523, 2012.

217. S. Sircar, "Influence of gas-solid heat transfer on rapid PSA," Adsorption, vol. 11, no. 1, pp. 509–513, 2005. ·

218. B. G. Keefer, A. Carel, B. Sellars, I. Shaw, and B. Larisch, "Adsorbent laminate structures," US patent 6, 692, 626, 2004.

219. B. G. Keefer and C. R. McLean, "High frequency rotary pressure swing adsorption: Apparatus," US patent 6, 056, 804, 2000.

220. A. S. T. Chiang and M. C. Hong, "Radial flow rapid pressure swing adsorption," Adsorption, vol. 1, no. 2, pp. 153–164, 1995. ·

221. J. Smolarek, F. W. Leavitt, J. J. Nowobilski, V. E. Bergsten, and J. H. Fassbaugh, "Radial bed vaccum/pressure swing adsorber vessel," US patent 5, 759, 242, 1998.

222. W. C. Huang and C. T. Chou, "Comparison of radial- and axial-flow rapid pressure swing adsorption processes," Industrial and Engineering Chemistry Research, vol. 42, no. 9, pp. 1998–2006, 2003.

223. D. M. Ruthven, "Past progress and future challenges in adsorption research," Industrial and Engineering Chemistry Research, vol. 39, no. 7, pp. 2127–2131, 2000.

224. P. Xiao, J. Zhang, P. Webley, G. Li, R. Singh, and R. Todd, "Capture of CO_2 from flue gas streams with zeolite 13X by vacuum-pressure swing adsorption," Adsorption, vol. 14, no. 4-5, pp. 575–582, 2008. ·

225. S. Dasgupta, N. Biswas, Aarti et al., "CO_2 recovery from mixtures with nitrogen in a vacuum swing adsorber using metal organic framework adsorbent: A Comparative Study," The International Journal of Greenhouse Gas Control, vol. 7, pp. 225–229, 2012.

226. Z. Liu, C. A. Grande, P. Li, J. Yu, and A. E. Rodrigues, "Multi-bed vacuum pressure swing adsorption for carbon dioxide capture from flue gases," Separation and Purification Technology, vol. 81, pp. 307–317, 2011.

227. C. Shen, Z. Liu, P. Li, and J. Yu, "Two-stage VPSA process for CO_2 capture from flue gas using activated carbon," Industrial & Engineering Chemistry Research, vol. 51, pp. 5011–5021, 2012.
228. C. A. Grande, F. Poplow, and A. E. Rodrigues, "Vacuum pressure swing adsorption to produce polymer-grade propylene," Separation Science and Technology, vol. 45, no. 9, pp. 1252–1259, 2010. ·
229. M. Tagliabue, D. Farrusseng, S. Valencia et al., "Natural gas treating by selective adsorption: material science and chemical engineering interplay," Chemical Engineering Journal, vol. 155, no. 3, pp. 553–566, 2009. ·
230. C. A. Grande and R. Blom, "Utilization of dual-PSA technology for natural gas upgrading and integrated CO_2 capture," Energy Procedia, vol. 26, pp. 2–14, 2012.
231. S. Sircar, "Separation of multicomponent gas mixtures," US patent, 4, 171, 206, 1978.
232. Y. Chen, A. Kapoor, and R. Ramachandran, "Two stage pressure swing adsorption process," US patent 5, 993, 517, 1999.
233. Y. Takamura, S. Narita, J. Aoki, S. Hironaka, and S. Uchida, "Evaluation of dual-bed pressure swing adsorption for CO_2 recovery from boiler exhaust gas," Separation and Purification Technology, vol. 24, no. 3, pp. 519–528, 2001. ·
234. D. L. Rarig, T. C. Golden, and E. L. Weist Jr., "Purification of CO_2 from H_2 PSA vent gas," inProceedings of the National AIChE Meeting, Indianapolis, Ind, USA, November 2002.
235. C. A. Grande and R. Blom, "Dual pressure swing adsorption units for gas separation and purification," Industrial & Engineering Chemistry Research, vol. 51, pp. 8695–8699, 2012.

Citations

CHAPTER 1

A. Lamberov, E. Dementyeva, D. Vavilov, O. Kuzmina, R. Gilmullin and E. Pavlova, "The Influence of Ceric Oxide on Phase Composition and Activity of Iron Oxide Catalysts," Advances in Chemical Engineering and Science, Vol. 2 No. 1, 2012, pp. 28-33. doi: 10.4236/aces.2012.21004.

CHAPTER 2

G. V. Figueroa Martinez, J. R. Parga Torres, J. L. Valenzuela García, G. C. Tiburcio Munive and G. González Zamarripa, "Kinetic Aspects of Gold and Silver Recovery in Cementation with Zinc Power and Electrocoagulation Iron Process," Advances in Chemical Engineering and Science, Vol. 2 No. 3, 2012, pp. 342-349. doi: 10.4236/aces.2012.23040.

CHAPTER 3

Sonali A. Borkhade, Preksha A. Shriwas, and Ganesh R. Kale, "Gasification Coupled Chemical Looping Combustion of Coal: A Thermodynamic Process Design Study," ISRN Chemical Engineering, vol. 2013, Article ID 565471, 11 pages, 2013. doi:10.1155/2013/565471.

CHAPTER 4

Marta Skolniak, Paweł Bukrejewski and Jarosław Frydrych (2015). Analysis of Changes in the Properties of Selected Chemical Compounds and Motor Fuels Taking Place During Oxidation Processes, Storage Stability of Fuels, Prof. Krzysztof Biernat (Ed.), ISBN: 978-953-51-1734-6, InTech, DOI: 10.5772/59805.

CHAPTER 5

Diab Mokeddem and Abdelhafid Khellaf, "Optimal Solutions of Multiproduct Batch Chemical Process Using Multiobjective Genetic Algorithm with Expert Decision System," Journal of Automated Methods and Management in Chemistry, vol. 2009, Article ID 927426, 9 pages, 2009. doi:10.1155/2009/927426.

CHAPTER 6

Nunzio Cennamo and Luigi Zeni (2014). Bio and Chemical Sensors Based on Surface Plasmon Resonance in a Plastic Optical Fiber, Optical Sensors - New Developments and Practical Applications, Dr Moh. Yasin (Ed.), ISBN: 978-953-51-1233-4, InTech, DOI: 10.5772/57148.

CHAPTER 7

Mariano Martín and Alberto Martínez, A Methodology for Simultaneous Process and Product Design in the Formulated Consumer Products Industry: the Case Study of the Detergent Business, doi.org/10.1016/j.cherd.2012.08.012.

CHAPTER 8

Carlos A. Grande, "Advances in Pressure Swing Adsorption for Gas Separation," ISRN Chemical Engineering, vol. 2012, Article ID 982934, 13 pages, 2012. doi:10.5402/2012/982934.

Index

A

Adsorption vi, 206, 207, 220, 221, 222, 223, 224, 225, 226, 228, 229, 230, 233, 234, 235, 236, 237, 239, 243
Attenuated total reflection (ATR) 76
Attenuated Total Reflection (ATR) 135

C

Carbon molecular sieves 216, 234
Chemical looping combustion (CLC) 35
Coherent scattering region (CSR) 3

Cyanide leaching 16

D

DCP (Direct-Current Plasma) 26
Diffuse reflectance infrared Fourier transformed spectroscopy (DRIFT) 77

E

Electrocoagulation (EC) 16, 18, 22
Electron Ionization Detector (EID) 72

F

Fatty acid methyl esters (FAME)

78, 90, 102

G

Gasifying agent-to-carbon ratio (GaCR) 40
Green rust (GR) 22

M

Mass Selective Detector (MSD) 72
Methyl t-butyl ether (MTBE) 78
Methyl tert-butyl ether (MTBE) 83, 84, 106, 107
Mixed-integer linear programming (MILP) 111
Molecular imprinted polymers (MIPs) 136
Molecular sieves 216, 222, 223, 235
Multiproduct batch 109, 110, 111, 113, 129

N

Nondominated Sorting Genetic Algorithm (NSGA) 112

O

Optimal design problem 109
Optimization problem 109, 110, 111, 112, 113, 119
Oxygen carrier (OC) 36

P

Parallel equipment 110, 113
Pressure swing adsorption 205, 221, 228, 229, 230
Processing stages 123

S

Scanning Electron Microscope (SEM/EDX) 27
Skarstrom cycle 209, 211
Surface Plasmon Resonance (SPR) 133
Surface Plasmon Wave (SPW) 135

X

X-ray phase analysis (XPA) 3